Y0-EGD-628

INDUSTRIAL ERGONOMICS

CASE STUDIES

INDUSTRIAL ERGONOMICS

CASE STUDIES

BABUR MUSTAFA PULAT AND DAVID C. ALEXANDER, EDITORS

McGraw-Hill, Inc.

New York St. Louis San Francisco Auckland Bogotá Caracas Lisbon London Madrid Mexico Milan Montreal New Delhi Paris San Juan São Paulo Singapore Sydney Tokyo Toronto

 Printed in the United States of America. Hardcover edition published by McGraw-Hill, Inc. and softcover edition published by the Institute of Industrial Engineers.

1 2 3 4 5 6 7 8 9 0 9 7 6 5 4 3 2

ISBN 0-07-050850-X

A note of thanks is extended to the following organizations for allowing us to reprint copyrighted material:

Figure 2-5. Cumulative Severity Rate appearing on page 25 is reprinted from *Human Factors*, Vol. 26, No. 6, 1984. ©1984 by The Human Factors Society, Inc. and reprinted with permission.

Table 11-1. Principles of Motion Economy for the Arrangement of the Work Station appearing on page 117 was adapted from B. W. Niebel, *Motion and Time Study*, 8th edition; Irwin Publishing Company, page 188.

Library of Congress Cataloging-in-Publication Data

Industrial ergonomics : case studies / Babur Mustafa Pulat and David C. Alexander, editors.

p. cm.

Includes bibliographical references.

ISBN 0-07-050850-X

1. Human engineering--Case studies. 2. Man-machine systems--Case studies. I. Pulat, Babur Mustafa. II. Alexander, David C., 1949- .

T59.7.I52 1992 91-43802

620.8'2--dc20 CIP

Table of Contents

Section I: Overview of Industrial Ergonomics

Section II: Physical Ergonomics

Section III: Information Ergonomics

Section IV: Design of Work Space and Work Methods

Section V: Product Design

Section VI: Macro-ergonomics

Section VII: Maintainability

Section I

Overview of Industrial Ergonomics

1

An Overview of Industrial Ergonomics

David C. Alexander, PE
B. Mustafa Pulat, PhD

THE NEED FOR INDUSTRIAL ERGONOMICS

What is Industrial Ergonomics?

Industrial ergonomics has come a long way in the past decade. Not long ago people were becoming acquainted with the terms *ergonomics* and *human factors*. Now they are learning about *industrial ergonomics*.

Industrial ergonomics deals with people in the workplace; specifically, it is the application of the sciences related to the study of work. *Ergon* means "work," and *nomics* is defined as "the study of" from the term *nomos*, meaning "laws." The modifier *industrial* may seem redundant, but its use further clarifies the application of ergonomics in the workplace or occupational setting. Because ergonomics can be applied to nonwork settings such as the home, automobiles, or consumer products, the modifier ensures the focus is on the industrial environment—manufacturing and service alike.

The need for industrial ergonomics is on the rise, because organizations are experiencing increased costs without it. Unions have negotiated clauses to implement ergonomics into their contracts, and the Occupational Safety and Health Administration (OSHA) has cited companies for failing to implement ergonomic solutions that will protect their employees. Designs that reflect proper ergonomic principles are advertised more frequently, because such designs are perceived as a marketing advantage. Managers as well as nonexempt workers are becoming familar with the term *ergonomics* and what it means.

Thus, the field of industrial ergonomics is growing rapidly, both in the number of practitioners and in the applications. The case studies presented in this book will highlight some of the more notable and innovative efforts. We hope these case studies will push the frontiers of industrial ergonomics further, by allowing others in the industrial environment to see what has been done and how it was accomplished.

An industrial ergonomics practitioner, called an *ergonomist*, studies workers and their tools, workplace, work space, environment, equipment, and facilities. By doing so, the ergonomist is able to create improved working conditions. Typical losses from the failure to apply ergonomics include [1]:

- Lower production output
- Increased lost time
- Higher medical and material costs
- Increased absenteeism
- Low-quality work
- Injuries, strains
- Increased probability of accidents and errors
- Increased labor turnover
- Less spare capacity to deal with emergencies

Ergonomics was once thought of as something that was nice to do. Now it is more often seen as something that is necessary to do. The contemporary business manager is learning that industrial ergonomics is cost effective and beneficial for the organization as well as for the workers, because it plays an important role in ensuring employee safety and health.

Costs Are Important

Today the costs of ignoring industrial ergonomics are significant. In October 1988, the meatpacking firm of John Morrell and Company was fined $4.33 million for a series of workplace injuries called *cumulative trauma disorders* (CTDs). This fine was a record amount for an OSHA citation, surpassing an earlier record fine imposed under similar circumstances on Iowa Beef Packers, another meatpacker.

Industrial ergonomics was instrumental in identifying and publicizing CTDs and later in developing the methods to correct these serious workplace injuries. The cost of one case of carpal tunnel syndrome, a common form of CTD, can reach $100,000; a cost of $40,000 per case is not unusual. In comparison, the costs to redesign workplaces are often far less expensive than even a single injury.

Ergonomics can also help reduce the high cost of worker's compensation. In 1985, for example, the cost for worker's compensation exceeded $22 billion, an increase of 14% since 1984. Back injuries accounted for 23% of the work injuries, and amounted to 32% of the total compensation costs. These were the highest for any of the part-of-body-injured categories. Relative to the type of accident, overexertion led the list with 30.7% of the cases. The average cost of an overexertion case was $4,750.[2]

Many overexertion cases are caused by lifting, pushing, pulling, and carrying objects. Ergonomists have studied these areas, and recommended measures to help prevent accidents and injuries. A study of the 10 leading work-related diseases and injuries in the United States revealed that five were directly or indirectly related to ergonomics.[3]

The cost of product liability is potentially large for any organization. Many ergonomics experts testify in cases where the product is allegedly defective from the ergonomic viewpoint. A major benefit of proper ergonomic design is fewer accidents and injuries and fewer opportunities for lawsuits, thus reducing the number of these cases. A firm that has a strong reputation for performing suitable ergonimcs work will have fewer lawsuits filed against it for negligent design.

Productivity and Quality Enhancements

Good ergonomic design can enhance productivity and quality, resulting in cost savings for organizations. In other words, failure to practice good ergonomic design can result in the failure to achieve an organization's potential.

This situation occurred frequently with the introduction of word processing equipment during the early 1980s. Word processing centers were justified based on studies that indicated substantial increases in productivity over the prior methods. Once installed, however, the productivity gains failed to materialize. The cause, demonstrated in many different situations, was due to poor ergonomic design. The neckaches, backaches, eyestrain, and hand-wrist problems that resulted from poor workplace design took away 15% to 30% of the expected productivity. Full gains were realized only when proper ergonomic workplace designs were implemented along with the new equipment.

Union Support

Unions are beginning to provide strong support for industrial ergonomics. The United Auto Workers (UAW) negotiated a contract with Ford Motor Company that resulted in substantial expenditures for ergonomic interventions at the Ford plants. The UAW published *Sprains & Strains: A Worker's Guide to Job Design*, a practical guide for the application of ergonomics.[4] The Amalgamated Textile Workers Union has published *A Worker's Guide to Job Design and Injury Prevention in the Clothing and Textile Industries*, a guidebook of ergonomic designs for their workplaces.[5]

Management Interest

Ergonomics is becoming known in business schools as part of the overview on worker safety and health. Articles about ergonomics are appearing in trade and technical journals. Of course, the OSHA citations and record fines are well-publicized in the popular press. With this information, responsible managers are asking about ergonomics and its impact on their operations. As a result, managers are becoming more responsive to their legal obligations, to the social needs and concerns of their employees, and to the economic realities of industrial ergonomics.

DESIGNS THAT INTEGRATE HUMANS INTO INDUSTRIAL SYSTEMS

Human Integrated Design

The essence of industrial ergonomics is to design with humans in mind. By considering the needs of humans from the initial design stages, they can be fully integrated with the other design elements. The term coined for this approach is *human integrated design* (HID). With human integrated design, the classic problems associated with a lack of ergonomics do not materialize.

When designing a system with humans in mind, the designer must consider the humans' role from the beginning of the design process and be aware of humans' capabilities and limitations. The designer must also consider three design aspects—hardware, operation variables, and human interaction with other systems (Figure 1-1).

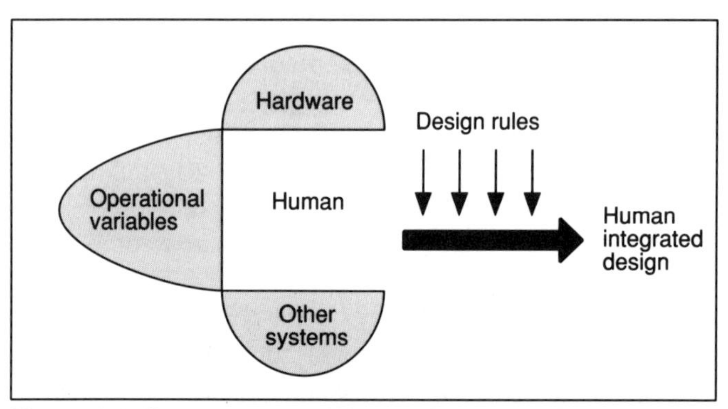

Figure 1-1. Components of Human Integrated Design.

The *hardware* aspect of the design are the things that can be touched or felt, such as conveyors, tables, chairs, control panels, keyboards, workbenches, and facilities. Because they are fixed elements in the design and are difficult and expensive to change later, the hardware should be designed correctly from the start. Changes and modifications after the initial design will be very expensive and result in inefficient design.

The *operation* or *software* aspect consists of many variables, including work methods, work pace, work-recovery regimens, and staffing levels. The operation aspect also includes the rules procedures, and activities associated with using the hardware. While this part of the system is dependent, in part, on the hardware aspect, it is also flexible to a degree, and can be modified after the hardware is in place. (In fact, the operation procedures are often modified to compensate for poorly designed hardware.) The entire design will work best, however, when the hardware and operation aspects are developed with humans in mind and when one aspect complements the other by supporting humans from the start.

In many other cases, employees interact with *other systems* as part of their work. For example, a receiving system at an incoming goods dock collects data related to material movement with a human operator's input. The compatibility of the operator's interface with this system and with expectations and prior use experiences is very important. Generally, people respond much better to user-friendly interfaces than to those that are user-hostile.

Principle 1: consider the human's role from the beginning. In order to consider humans from the beginning, the designer must appreciate the humans' role in the final operation of the system. Principle 1 is true whether the design is for a new aircraft, a kitchen, or a handtool. Once the designer perceives that humans will play a key role in the successful use of the system, the emphasis on considering humans early in the design stage can follow. Failure to do so will result in compromise and suboptimal use of the system later on.

A human's role in a system can take many different forms. At one extreme, the human may play an integral role in the use of the system. A common example is the automobile, where the human is the controller, the passenger, and the provider of service and maintenance. The automobile is of no use without the human. Many traditional industrial systems are like this; the operation of machine tools, the control of processes, and the manning of equipment is traditionally done by people. The use of numerical control, automatic process controllers, and now, robotics and advanced manufacturing systems has and will continue to alter the control aspects of these industrial activities. People, however, will continue to play a strong role in most industrial systems for many years to come.

The newer industrial systems bring up the other end of the spectrum involving people. In robotic systems, for example, humans initially appear not to be involved and that there is little need for HID. Appearances, as they say, are deceiving. People are involved in programming the robot, in service, and in maintenance; therefore, those aspects of the system must be designed so that humans can perform those functions effectively.

When programming is done from a keyboard, as with the initial robots, the programmer is less effective in conceptualizing all the spatial relationships and developing programs. A better programming design is to use the teaching pendant that allows the programmer to manipulate the robot arm through a series of moves, which the robot will eventually mimic and duplicate endlessly.

The widespread use of the teaching pendant is testimony of the power of human integrated design. In these newer designs, it only appears that humans are not involved; the involvement occurs at a different level. Because the human programmers are still critical, they must be considered before the design can be fully effective.

Principle 2: be aware of the human's limitations and capabilities. Principle 2 proposes that the designer must be aware that any type of design requires a detailed knowledge of what is involved in that design. This is true whether the designer works with steel I-beams and similar structural components or with people and other parts of a complex system.

The designer also needs information about the humans' limitations and capabilities of the humans involved in the system or design. The designer must know what the humans cannot do or cannot do very well. At the same time, the designer must know what the humans can and will do well, and therefore should be doing. By understanding such strengths and limitations, the designer can decide which duties to allocate to the humans and which to give to the other components of the system.

This is not to say that the allocation of functions within a complex system involves nothing more. In many cases, for example, there is no allocation decision to be made. The machinery simply cannot perform a necessary function. Consequently, a default decision requires that a human perform the function even if that particular human is not fully capable of the task either. In other cases, a function is assigned to both, one complementing the other.

Automation, on the other hand, can be the answer to dull, dirty, difficult, demanding, and highly repetitive tasks. Automation can replace human performance once the capability to perform the task in question and outperform humans from the standpoint of effectiveness, efficiency, or cost is achieved. The realization that a task can be performed better without humans than with humans is sometimes slow in coming, but it can result in workplace innovation. Human operators, however, have to perform the task until automation can. An example is the use of bar-coded sorting systems that use a laser scanner to identify products, and then control a mechanical system to sort and route them. People operated these systems for many years prior to the advent of the bar-coding technology.

Other aspects. Human integrated design involves several other important factors, including the cost of the design, new design versus evolutionary design, and the cost of implementing the design. The cost of a new design is always a powerful consideration. Any veteran designer is well aware of the need to control design costs. Budget and time constraints placed on design work are often used as excuses for not fully considering the humans in the early stages.

The managers who make key funding decisions must be aware of the concept of fully costing each design. Full cost includes the cost of designing and building the item and the cost of operating it throughout its life cycle. The product that is designed and made inexpensively but is expensive to operate is often less of a bargain. Small investments at the design stage can pay off repeatedly. On the other hand, shortcuts in design can result in small savings. The cost comes later during less than optimal operating conditions.

Unfortunately, the designer is rarely responsible for operations. The designer may see the incentives to shortcut the design process and none of the penalties. In other words, the designer can shortcut the design process, fail to fully integrate the human aspect, and still never understand that a particular design is deficient. The user, however, will know that using the product is difficult, time consuming, and awkward but will be powerless to make any fundamental changes that will improve the situation. Higher than expected operating costs, increased errors, and worker injuries are the result. Figure 1-2 illustrates the relative cost differences between operating and design costs.

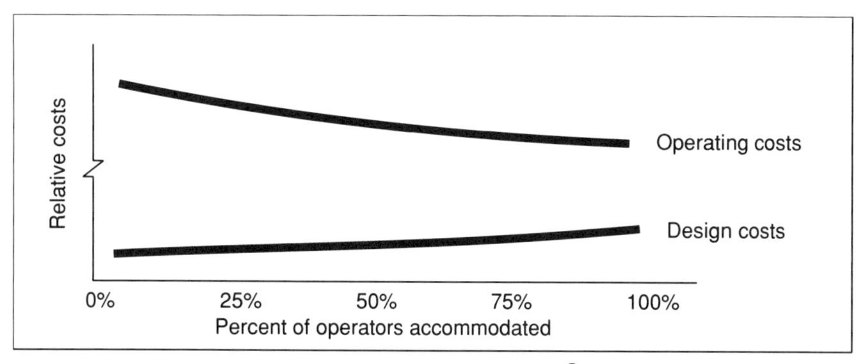

Figure 1-2. Design and Operating Cost Trade-offs.[3]

The issue of new design versus evolutionary design plays a strong part in human integrated design. In new designs, the full role of humans may be difficult to identify and anticipate. With evolutionary designs, it will be possible to determine the problems with the prior design and to develop corrections. Our view is that the more human integration that is carried out initially, the better.

Also at issue is the cost of implementing the design. Even if there is a proper HID, it may not be implemented because it may be too expensive to fully do so. Given budget limitations, the managers responsible for implementation may choose a less expensive design or may omit critical components of the human integrated design. This approach may prove to be more costly later on. For example, while a robot may be the preferred alternative to avoid placing the employee in a demanding and dangerous job, the cost may be prohibitive. The result is that the employee is still placed in a suboptimal design.

Human Integrated Manufacturing

Human integrated design leads to human integrated manufacturing (HIM) in the manufacturing environment (Figure 1-3). A principle concern is whether human operators can effectively work with the design. Another concern is whether or not the design documents, instructions, and other design support issues are adequate.

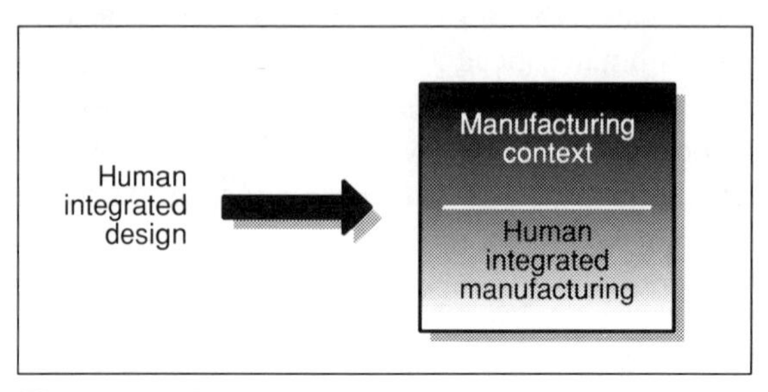

Figure 1-3. The Relationships of Human Integrated Design and Human Integrated Manufacturing.

Two examples illustrate the second concern. The first involves design documents from a product design organization that may be so difficult to read that a bill of material (BOM) analyst may order incorrect parts or may order incorrect quantities of the correct parts. Then, everyone in manufacturing wonders why there are both part shortages and overages. The second example involves assembly line operators who are unable to correctly read engineering information from a video display terminal, which is presenting glare, orientation, and resolution problems. As a result, components are inserted into the wrong holes on a printed wiring board. These human errors, which are a direct result of design flaws, can lead to thousands and sometimes millions of wasted dollars.

Human integrated manufacturing is a philosophy, not a technology. It must be exercised at every stage of manufacturing from design through material procurement, buffering, assembly and test, and shipping. People must always be thought of as an integral part of every manufacturing system, even though their extent of involvement may vary from system to system. Remember, people are the only system element that can plan, think, and act according to the plan. They can make or break a system.

THE PRIMARY AREAS OF INVOLVEMENT FOR INDUSTRIAL ERGONOMICS

When dealing with people in industrial systems, there are several applications to consider, each of which deals with a different aspect of human performance.

Although summarized here, these applications are discussed in detail in various chapters throughout this book.

One application is *physical ergonomics*, which deals with the physical well-being of humans. The objective is to minimize musculoskeletal disorders due to single-incident or repetitive trauma. Evaluating the loads and anticipating problems, redesigning jobs, and selecting and training workers are several methods of dealing with real or potential physical ergonomics problems.

Information ergonomics examines the way in which humans obtain and processes information. The proper design of all elements and interrelationships of these elements on such tasks can reduce human error and make these jobs safer and more effective.

The *design of the work space and work methods* is yet another involvement area for industrial ergonomics. By examining where, how, and why workers work, the resulting operator-machine system can be improved. The system can be made more productive and still provide a safe, healthy work environment.

Product design is another important area of industrial ergonomics. Designing products so that they are compatible with human-use patterns without inflicting injury to users makes the products more effective, allows users to enjoy them, and often enhances learning and skill with the product.

Macroergonomics, a relatively new application of industrial ergonomics, is concerned with the employees' whole job performance, including the motivational aspects of the job and the workers.

Designing systems for maintainability and reliability are also important. New areas of application include workers in advanced manufacturing systems. Because many tasks are being automated, the workers' role continues to be shifted from a force generator to a monitor of tools and equipment. Consequently, mental load is increasing. An industrial ergonomist must be able to design such interfaces so that employees can still function effectively in such systems. This is another case of human integrated design and manufacturing.

SUMMARY

As a concept, ergonomics is growing fast in the industrial environment. Both the federal and state governments and the private sector are realizing the benefits of implementing projects that integrate people into the design. Catalysts in this process include the practicing ergonomists, research ergonomists, labor unions, and sound business practices.

An organization must be able to position itself for enhanced employee relations and productivity with the aid of ergonomics. This may be achieved through educational programs and implementation of key projects. Every engineer and supervisor, along with others who have design responsibilities, must have knowledge in this area. Once the foundation is laid, success with ergonomics will follow.

References

1. Alexander DC, Pulat BM, ed. *Industrial Ergonomics: A Practitioner's Guide.* Norcross, GA: Industrial Engineering and Management Press; 1985.

2. *National Safety Council Accident Facts.* Chicago, IL: National Safety Council; 1988.

3. Alexander DC. *The Practice and Management of Industrial Ergonomics.* Englewood Cliffs, NJ: Prentice-Hall, Inc.; 1986.

4. MacLeod D. *Strains & Sprains: A Worker's Guide to Job Design.* Detroit, MI: International Union, United Automobile, Aerospace, and Agricultural Implement Workers of America, UAW; 1982.

5. *Worker's Guide to Job Design and Injury Prevention in the Clothing and Textile Industries.* New York, NY: Amalgamated Clothing and Textile Workers Union, AFL-CIO; 1988.

Section II

Physical Ergonomics

2

JSI Method for the Control of Manual Materials Handling Injuries

M. M. Ayoub, PhD

Although many machines used in industry are capable of replacing manual work, a completely automated world—a world free of manual work—is still far beyond reality. Either economic considerations such as the cost of automation equipment, or practical situations such as space limitations or unexpected conditions, can sometimes make complete automation difficult. Consequently, manual work is unavoidable in many workplaces.

Manual work, whether heavy or light, can lead to overexertion and repetitive trauma disorders. Overexertion in manual materials handling (MMH), for example, is a major cause of back injuries,[1] and the corresponding expenses and losses are overwhelming.

Figure 2-1 illustrates the frequency and cost of overexertion injuries in the United States. Efforts through industrial ergonomics and other applications have reduced the number of injuries in the last several years, but costs continue to escalate. Manual materials handling activities produced the single largest percentage of compensable work injuries in the United States, accounting for one-fourth to one-third of all injuries.[2] The cost to US businesses is estimated at $4 billion to $20 billion annually.[1,3]

This chapter provides an historic perspective of one type of MMH injury, cumulative trauma disorders, and introduces the Job Severity Index (JSI), an effective approach for evaluating physically demanding lifting jobs to control manual materials handling. The goal of the JSI is to control MMH injuries before the manufacturing process is finalized.

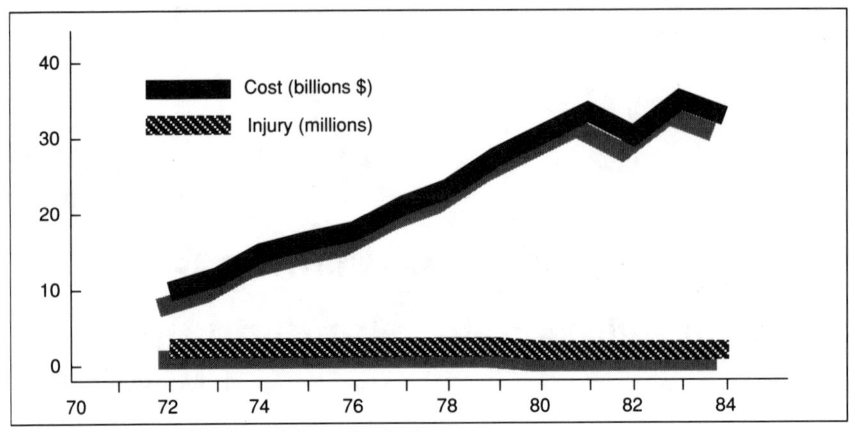

Figure 2-1. Work Injury and Cost Statistics.

HISTORIC PERSPECTIVE

Until recently, *cumulative trauma disorders* (CTDs), also referred to as *microtrauma*, *repeated trauma*, or *chronic trauma*[4] have commanded relatively little attention from the public as well as from ergonomists. Yet, early medical reports provide adequate evidence that over the years skilled craftsmen have experienced a variety of musculoskeletal disorders related to their trades. Names, such as brick layer's shoulder, carpenter's elbow, stitcher's wrist, game keeper's thumb, telegraphist's cramp, and cotton twister's hand, represent a sampling of the labels for such syndromes.[5] Industry is beginning to address the adverse health effects that result from CTDs and to better understand the affects of repetitive trauma injuries, irregular motions, and postures.

Improvement Approach

Industrial and technologic advances in the United States have been greatly motivated by the need to overcome and solve design, production, and distribution problems of manufactured products. Traditionally, industry subscribed to the *improvement approach* as the means of becoming more economically efficient. Using this approach, products are often designed and developed rather quickly and then a manufacturing process is determined.

Over a period of time, the deficiencies in the product design and the manufacturing process are removed during the production process. Improvements to the products and the manufacturing process can take several years, and can lead to higher manufacturing expenses coupled with a greater risk of worker injury and higher medical costs. This approach also can lead to possible retributions by regulatory agencies, such as the Occupational Safety and Health Administration (OSHA).[6] Regardless of these considerations, the improvement approach to product design and manufacturing is still used in many industries.

Industry must take the initiative of designing good products and manufacturing processes to reduce health and safety problems in the workplace. To do this, ergonomists must be able to show that the maintenance of occupational health and safety and the maintenance of high levels of productivity go hand in hand.

The usual pattern, however, is that management's attention does not focus on ergonomics until there is a breakdown in occupational health and safety during an established productive operation. Under such circumstances, past activities and events are reviewed and evaluated. This review and evaluation uses the four Cs of occupational safety and health investigations—cause, consequence, cost, and cure.[6] In most situations, consequence and cost are generally known, but the cause and the cure must be discovered by ergonomists and other professionals.

Theoretical analysis is one way to resolve some of these problems, as is the implementation of experimental methods for evaluating alternative solutions. Also, industrial ergonomics includes several broad branches for problem solving, including biomechanics, work physiology, and engineering psychology. These branches combine to cover problems ranging from physiologic loads and biomechanical problems to mental workload and related stress.

Ergonomists view the work situation as a system composed of three components: the operator, as the human element in the system; the task that needs to be performed with tools and equipment as needed; and the environment in which the work must be performed. These work system components produce a complex interaction that produces work stress, which in turn results in work strain, affecting not only worker satisfaction but also injuries, productivity, and associated costs (Figure 2-2).

JOB SEVERITY INDEX

Generally, as the load lifted increases, the probability of injury increases. Injury data collected in industry showed that some demanding lifting activities did not always result in higher injury rates.[7] Rather, less demanding tasks had significantly higher injury areas than those found in demanding tasks, which may be attributed to the ability of the operator assigned to the job. These results yielded a hypothesis that injury rates are expressed as a function of two variables:

Injury Rates (Injuries per 200,000 Hours Worked) =
Functions of (Task Demands, Operator Capacity)

Consequently, the JSI was proposed to assist in the design and identification of jobs that may be demanding enough to require ergonomic evaluation and modification.

The JSI reflects both job demands and operator capacity. For such an index to be meaningful, it must relate to the frequency and severity of injuries as the numeric value of the index increases. Conceptually, the JSI is represented by the equation:

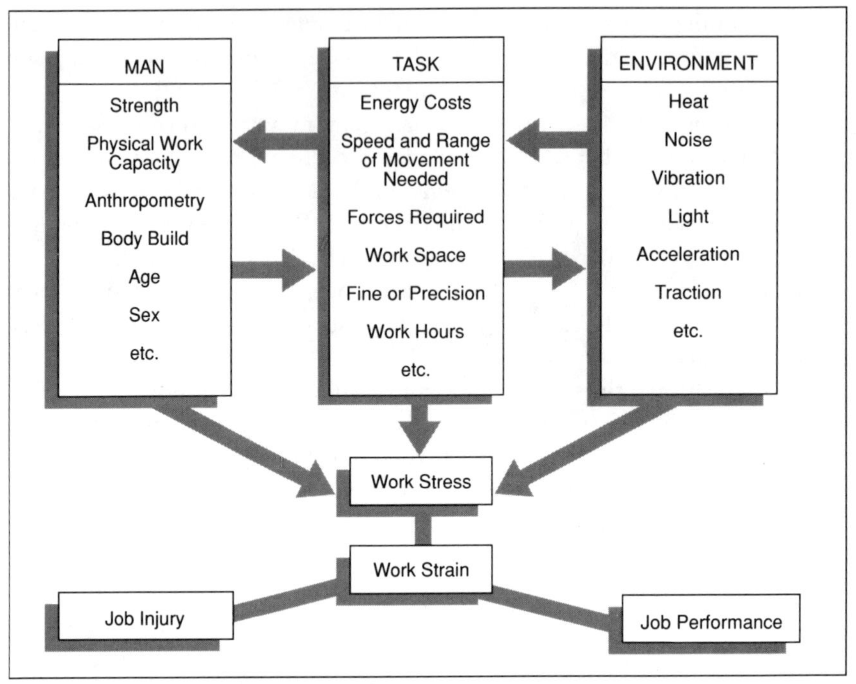

Figure 2-2. Man-Task-Environment System.

$$\text{JSI} = \frac{\text{Job Demand}}{\text{Worker Capacity}}$$

In order to quantify JSI, job demands and operator capacities must be specified for the job conditions. These include container size, starting and ending points of the lift, and frequency of the lift.

Since most operators do not work on the same job or on the same task every day for the entire shift, the JSI equation can be modified to reflect the fact that several tasks can be performed, all of which differ in terms of job conditions. This modification is represented by the following equation:

$$\text{JSI} = \sum_{i=1}^{N}\left[\frac{\text{hours}_i}{\text{hours}_t} \times \frac{\text{days}_i}{\text{days}_t}\right] \times \sum_{j=1}^{M_i}\left[\frac{F_j}{F_i} \times \frac{WT_j}{CAP_j}\right]$$

where

N = number of task groups
hours_i = exposure hours/day for Group i
days_i = exposure days/week for Group i
hours_t = total hours/day for job
days_t = total days/week for job

M_i = number of tasks in Group i
WT_j = maximum required weight of lift for Task j
CAP_j = the adjusted capacity of the person working at Task j
F_j = lifting frequency for Task j
F_i = total lifting frequency for Group i

$$= \sum_{j=i}^{M_i} F_j$$

The index is also the function of the ratio of job demands to lifting capacity of the operator(s) performing the job, and is conceptually similar to the lifting strength ratio.[8] In the JSI relationship, job demands include determinable quantities through comprehensive task analysis, such as weight and frequency of lift and the task geometry.

Worker capacity, on the other hand, is an estimated quantity that is a function of measurable human characteristics, including strength, body size, and endurance measures. The worker capacity used in the job severity index equation is an estimated maximum acceptable weight of lift (MAWL) derived from equations developed by Ayoub *et al.*[9] A set of mathematical models to predict the MAWL has been developed from the results of a large industrial study.[9,10] This study asked 73 men and 73 women to determine their maximum acceptable weight of lift under various conditions, including lifting frequency, container size, and lifting range.

During the MAWL data collection, strength, anthropometric, and endurance data were also obtained on the 146 subjects. Regression equations were developed to estimate an individual's lifting ability at a given set of job conditions based on that individual's known measurements of strength, anthropometry, and endurance.[9,10]

The JSI method requires that each job be described in detail as a series of lifting activities or tasks. A *lifting task* is defined as a unique movement of an item from one point in space to another. Each task is then described in terms of the maximum required weight of lift; the largest required load center of gravity at lift initiation and at lift termination; the load height (vertical distance from the floor to the hands) at lift initiation and at lift termination; and the task exposure time in hours. A sample job description is presented in Table 2-1.

Each task is grouped with others performed during the same time period, which means that some jobs require the performance of different sets of tasks during different times of the day. Describing the job as a series of lifting tasks and grouping those tasks is necessary to properly account for the relative importance of each job component. In addition, the lifting range is determined by using information[9] that divides the lifting ranges into six categories:

1. Floor to knuckle
2. Floor to shoulder
3. Floor to reach
4. Knuckle to shoulder
5. Knuckle to reach
6. Shoulder to reach

Table 2-1. Job Description Information.

Task 1	Max weight = 22 Lifting frequency = 0.0500 Max box size = 12 Load height (initial) = 23 to 23 Load height (terminal) = 47 to 58 Task hours/days = 8.0/5.0
Task 2	Max weight = 25 Lifting frequency = 0.1500 Max box size = 11 Load height (initial) = 9 to 64 Load height (terminal) = 35 to 35 Task hours/days = 8.0/5.0
Task 3	Max weight = 2 Lifting frequency = 0.0333 Max box size = 6 Load height (initial) = 4 to 42 Load height (terminal) = 47 to 47 Task hours/days = 8.0/5.0
Task 4	Max weight = 6 Lifting frequency = 0.500 Max box size = 11 Load height (initial) = 35 to 35 Load height (terminal) = 9 to 64 Task hours/days = 8.0/5.0
Task 5	Max weight = 6 Lifting frequency = 0.1667 Max box size = 6 Load height (initial) = 9 to 64 Load height (terminal) 47 to 47 Task hours/days = 8.0/5.0
Task 6	Max weight = 50 Lifting frequency = 0.0500 Max box size = 3 Load height (initial) = 41 to 41 Load height (terminal) = 58 to 58 Task hours/days = 8.0/5.0

Because a task, such as loading a pallet, may require lifting over several different ranges, capacity calculations are performed for each of the required lifting ranges. For JSI to be conservative, the smallest of these capacities represents the entire task; therefore, JSI values greater than one are not uncommon.

The JSI equation includes frequency corrections that are based on the total lifting frequency for the task group to which the task belongs. Task group lifting frequency is the sum of the lifting frequencies of all tasks in a task group. The center of gravity correction is based on the larger of the two task centers of gravity: lift initiation and lift termination. Task center of gravity is the horizontal distance between the workers' ankles and the center of gravity of the item being lifted. The adjusted capacity is simply the predicted capacity less body weight, multiplied by the frequency correction factor and the center of gravity correction factor. The center of gravity correction factor is based on container size.[9]

The next step is to divide the maximum required weight of lift at each task by the adjusted capacity of the workers assigned to each task, which is the smallest capacity of the task required. These individual task ratios are then weighted according to the lifting frequency and averaged with other tasks in the same group. The final step in the JSI calculation is to take an exposure time-weighted average of the task group ratios. The result is the job severity index for that job.

DATA COLLECTION AND USE OF JSI

Two large field studies required an accumulation of a large amount of data on job requirements, worker capacities, and injuries over a period of time.[11,12] Both studies required the participation of organizations, in which employees perform manual materials handling. Within each of the companies involved, only those jobs requiring regular and frequent lifting were selected for this study.

A regular job, one that has requirements that remain relatively constant over time, was selected to avoid bias due to job changes that could affect the injury data over time. A frequent lifting job requires at least 25 lifts per day of not less than 5 kg and requires an exposure of at least two hours per day. Most of the jobs selected exceeded the minimum requirements. Each selected job was analyzed using the JSI method by direct observation, by inspection of production records, and by direct interview of workers and supervisors to determine the job requirement data. A total of 101 jobs representing a wide range of different manual material handling were studied at the various plants.

Those who were chosen from the workers performing each selected job met certain minimum requirements, including at least six months experience on the job and no serious health problems. After all the workers were selected, anthropometric measurements and strength and endurance tests were performed. A comprehensive injury profile was also compiled on each worker beginning immediately after the worker measurement was made and continuing until the worker changed jobs, the job was changed, or the studies ended. This time period ranged from one month to more than two years.

The profiles contained information about all injuries sustained by each worker during the study period regardless of injury cause, type, or severity. The injury data, generally classified as either a lifting injury or a nonlifting injury, were obtained through inspection of company personnel files, from insurance carriers, or in a few cases, by interviewing the worker's supervisor. The lifting injury information gathered for the two studies included injury type and was divided into the following classes:

- Musculoskeletal injuries to the back
- Musculoskeletal injury to other body parts
- Surface tissue injury due to impact
- Other surface tissue injury
- Miscellaneous injury

Table 2-2 provides a summary of injury data collected for the lifting injuries.

ANALYSIS AND RESULTS

The relationship between worker JSI and injury is important when considering such an approach's validity as a tool for employee screening and placement, and for job design and redesign. Theoretically, if JSI evaluations were done for each person seeking assignment to a specific job and if only those workers with acceptable JSI values were assigned to the job, then the injury rates on that job should also be acceptable. This assumes, of course, that a low JSI value implies a low injury potential. The results thus far support this assumption.

In order to use JSI as a screening tool, individual JSIs would have to be calculated for each worker. In these studies, such calculations required the job requirements data for the job in which each worker was working and the subjects' measurement data, as needed to predict capacity. When this was completed, the results showed a set of 453 JSI values, one for each of the 244 workers in the first study and one for each of the 209 workers in the second study.

The workers were ranked according to individual JSI values and divided into 10 JSI groups, ranging from very low severity to very high severity levels. The injury data for the workers in each group were calculated. Each group represented approximately the same number of hours of worker exposure; therefore, the injury statistics should be comparable.

The results are based on the injury experiences of 453 workers in 101 different industrial lifting jobs, representing a total of 1,057,088 hours of worker exposure. The *injury rate statistic* used in these studies is defined as the total number of lifting back injuries per 100 full-time employees (FTE). The *disabling injury rate statistic* is defined as the number of disabling back injuries (one or more lost work days) per 100 FTE. The *severity rate statistic* is defined as the number of days lost per disabling injury.

The injury rate, disabling injury rate, and severity rate are presented in Figures 2-3, 2-4, and 2-5. These figures show cumulative injury statistics for

Table 2-2. Injury Data Collected*.

	Injury Cause and Type†				
	L1	L2	L3	L4	L5
Number of injuries	23	26	38	1	0
Number of lost-time injuries	15	8	10	0	0
Number of days lost	693	22	20	0	0
Number of injuries	32	17	32	1	8
Number of lost-time injuries	21	9	6	0	3
Number of days lost	453	67	34	0	8
Medical expenses ($)	29,627	3,166	2,417	25	487
Wages paid during lost days ($)	3,366	1,115	1,033	0	385
Workers' compensation paid ($)	28,334	6,485	368	0	0
Total expenses ($)	**61,327**	**10,766**	**3,818**	**25**	**872**
	Injury Cause and Type†				
	N1	N2	N3	N4	N5
Number of injuries	3	33	47	3	33
Number of lost-time injuries	2	10	9	1	7
Number of days lost	10	163	15	1	7
Number of injuries	5	16	72	0	6
Number of lost-time injuries	2	4	21	0	1
Number of days lost	6	62	251	0	1
Medical expenses ($)	597	811	13,768	0	165
Wages paid during lost days ($)	220	1,560	2,678	0	0
Workers' compensation paid ($)	0	1,403	7,753	0	0
Total expenses ($)	**817**	**3,774**	**24,199**	**0**	**165**

*Study 1: 244 subjects, 568,250 exposure hours
Study 2: 209 subjects, 489,631 exposure hours
†Injury cause: L=lifting, N-nonlifting. Injury type: 1=musculoskeletal injuries to the back, 2=musculoskeletal injuries to other body parts, 3=surface tissue injury due to impact, 4=other surface tissue injury, 5=miscellaneous injury.

segments of the work force (453 workers) performing at or below various JSI levels. Figures 2-3, 2-4, and 2-5 also show the existence of injury threshold at a JSI of approximately 1.5.

This hypothesis, originally based on the results of the first study, was also supported independently by the results of the second study. It was further supported by the cost data collected during the second study only. The total direct injury expense for those working at JSI levels above 1.5 was approximately $60,000 per FTE as compared with an injury expense of only $1,000 per FTE for those people working at a JSI value of 1.5 or less.[10]

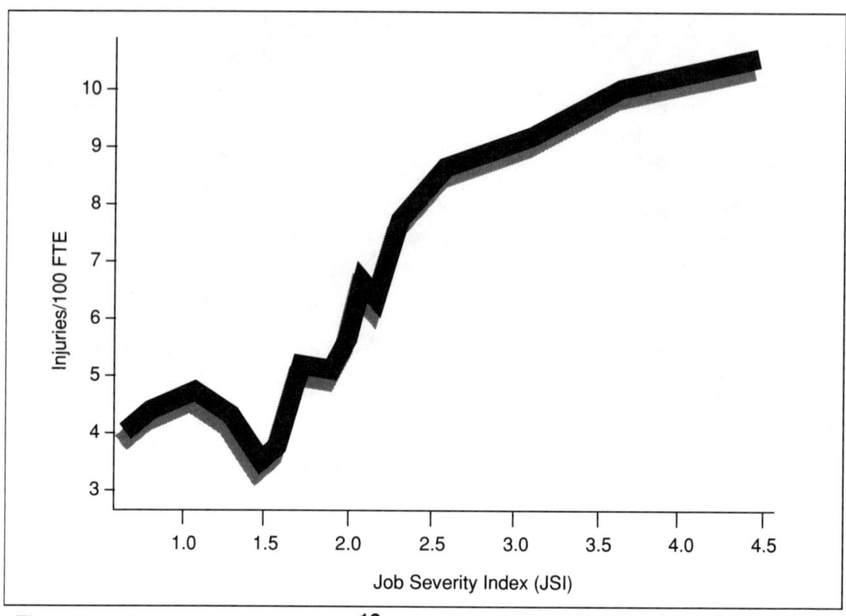

Figure 2-3. Cumulative Injury Rate.[10]

While JSI has applications for employee placement, it has strong implications for job redesign. Theoretically, if a given job was designed or redesigned so that a small percentage of workers would be overstressed, then a comparatively low level of injury should be expected. This is assuming that a low percentage overstressed implies a low injury potential. Results of these studies support such a hypothesis.

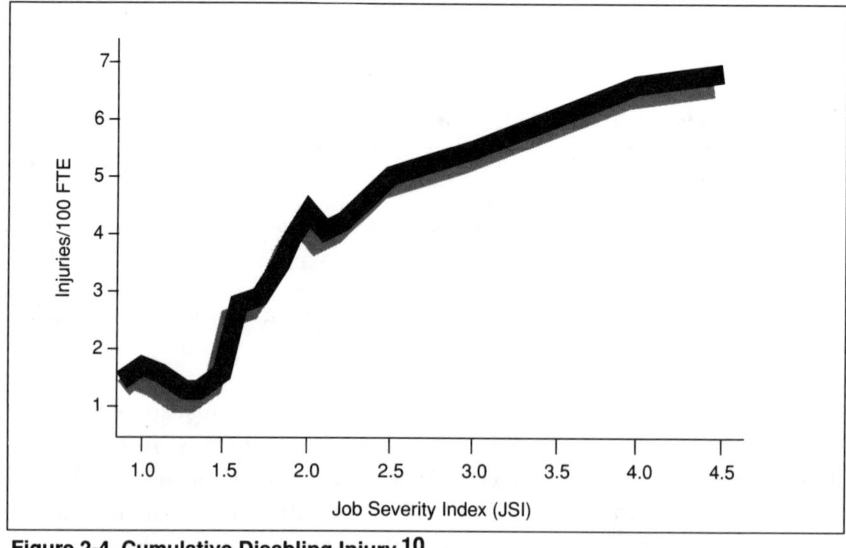

Figure 2-4. Cumulative Disabling Injury.[10]

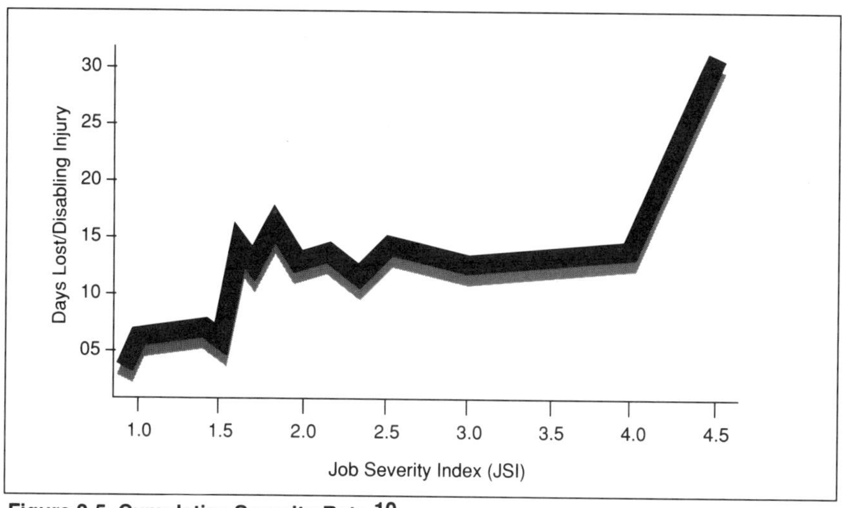

Figure 2-5. Cumulative Severity Rate.[10]

Using the data, the first step in the design analysis was to determine the amount of stress each job presented. This was done by predicting the percentage of the entire work force that would be overstressed if working in a particular job. The work force was represented in these calculations by a small sample of 385 workers, which consisted of all the male workers measured in both studies. Using the predicted capacities of the male workers, 385 calculations were performed for each job. That proportion of the sample with JSI values greater than 1.5 was defined as the percentage overstressed for the job.

The jobs were then divided into one of three stress categories. The first category included those jobs that overstressed 5% or less of worker population used in the study; the second category included those jobs that overstressed more than 5% but less than or equal to 75% of the sample population; while the third category included those jobs that overstressed more than 75% of the sample population. The categories are similar to those used in the work practices guide of the National Institute of Occupational Safety and Health.[1]

This job classification procedure did not consider the female workers because of the relatively small number of women in the subject population on these jobs. For comparison, however, the data seemed to indicate that a job that overstresses 5% of the male workers would overstress approximately 50% of the female workers, and one that overstresses 75% of the male workers would overstress virtually all of the female workers.

Based on the information, cost statistics were calculated for each job stress category. These statistics, presented in Table 2-3, show a definite increase in injury rates and costs as the percentage overstressed increase. The days-lost statistic, however, was unexpected because of a worker with a small predicted capacity who was seriously injured in a relatively low-stress job. This illustrates the point that even if a job accommodates a large percentage of the work force, injuries may be sustained by workers who are overstressed due to their small capacity. These job design results have been compared with results from a similar analysis based upon the methods discussed in the work practices guide.[1] This comparison indicates that the two methods yield similar results.[11]

Table 2-3. Injury and Cost Observed in Various Job Severity Categories.

	Percentage of Population Overstressed		
	$\% \le 5$	$5 < \% \le 75$	$\% > 75$
Number of Hours	305,333	510,485	342,063
Injury Rate*	5.24	10.97	15.70
Disabling Rate*	1.75	15.11	51.07
Expense Rate*	9,208+	35,092	36,337

*Per 100 FTE
+Cost data for second study only

CONCLUSIONS AND RECOMMENDATIONS

The job severity index is a valid and effective tool for controlling manual materials handling injuries. Although the JSI method seem to be feasible for both job design and employee selection, a question may arise as to why the threshold does not appear at a JSI of 10 but rather at 1.5. The reason is that in the JSI method the basic ratio is equal to the maximum appropriate predicted capacity. This is the worst-case procedure and, therefore, is conservative.

The primary focus of any injury control attempt using this method should be job design or redesign; employee selection should be secondary. There are two reasons for this statement:

1. Job design modifications are relatively permanent and require no ongoing programs. Employee selection, on the other hand, requires that each person be measured and evaluated for placement.
2. Employee screening and selection techniques of any type are difficult to apply because of the constraints imposed by government and organized labor.

This method may favor large, strong men over small, less strong women. Although this apparent discrimination is, of course, based on measurable human characteristics and solid experimental observations, it is difficult to guarantee that a specific woman would be injured more frequently than a specific man when working on the same job.

Procedures such as JSI are a good example of designing programs to help ergonomists design better workplaces and manufacturing processes. The JSI is specifically applicable for manual materials handling and lifting in particular. Similar approaches can be used in the design of other tasks.

The important lesson to be learned here is that a complete task analysis must be performed prior to implementation in the manufacturing process. Based on that complete task analysis, tasks must be designed so that the stress imposed on the workers is minimized due to the job's physical demands. If this can be accomplished, then the hazards at the workplace will be minimized while maintaining or improving productivity and reducing costs.

References

1. National Institute for Occupational Safety and Health. *A Work Practices Guide for Manual Lifting.* Cincinnati, OH: U.S. Department of Health and Human Services (NIOSH); 1981; Technical Report No. 81-122.

2. Kroemer, KHE. An isoinertial technique to assess individual lifting capability. *Human Factors.* 1983;25(5):493-506A.

3. Norby EJ. Epidemiology and diagnosis in low back injury. *Occupational Health and Safety*; 1981:38-42.

4. Anderson VP, ed. *Cumulative Trauma Disorders: A Manual for Musculoskeletal Disease of the Upper Limb.* London: Taylor and Francis Ltd.; 1988.

5. Hunter D. *The Disease of Occupations*, 6th ed. London: Hodder and Stoughton; 1978:857-864.

6. Tichauer ER. *The Biomechanical Basis of Ergonomics: Anatomy Applied to the Design of Work Situations.* New York, NY: Wiley Interscience; 1978.

7. Ayoub MM, Selan JL, and Liles DH. An ergonomics approach for the design of manual materials handling tasks. *Human Factors*; 1983;25:507-515.

8. Chaffin DB and Park KS. A longitudinal study of low back pains as associated with occupational weight lifting factors. *American Industrial Hygiene Association Journal.* 1973;34:513-525.

9. Ayoub MM, Dethea NJ, Deivanayagam S, et al. *Determination and Modeling of Lifting Capability, Final Report.* Lubbock, TX: Texas Tech University; 1978; NIOSH Grant No. 5R010H00545-02.

10. Liles DH, Deivanayagam S, Ayoub MM, and Mahajan P. A job severity index for the evaluation and control of lifting injury. *Human Factors.* 1984; 26:683-694.

11. Liles DH, Mahajan P, and Ayoub MM. An evaluation of two methods for the injury risk assessment of lifting jobs. In *Proceedings of the Human Factors Society 27th Annual Meeting.* Santa Monica, CA: Human Factors Society; 1983:279-283.

12. Ayoub MM, Liles DH, Asfour, SS, et al. *Effects of Task Variables on Lifting Capacity, Final Report.* Cincinnati, OH: National Institute for Occupational Safety and Health;

M. M. Ayoub is a P. W. Horn Professor of Industrial Engineering. He holds a BS in Mechanical Engineering from the University of Cairo, and MS and PhD degrees in Industrial Engineering from the State University of Iowa and is a registered Professional

Engineer in Texas. Dr. Ayoub's special interests lie in the areas of biomechanical modeling of movement and modeling of stresses imposed on the body while performing demanding tasks.

Dr. Ayoub has published over 100 articles in refereed journals and authored chapters in two recent books on human factors. He also serves on the editorial boards of *Industrial Ergonomics* and the *Journal of Safety Research*. He is a member of IIE, Human Factors Society, Ergonomic Society, American Society of Biomechanics, AIHA, and the American Society of Engineering Education.

3

Design and Analysis of Multiple Activity Manual Materials Handling Tasks

Anil Mital, PhD

Manual handling of materials (MMH) is regarded as the primary source of overexertion injuries and low back pain in industry, with one out of every three or four industrial overexertion injuries attributed to MMH. These injuries generally include shoulder and hip injuries and lower back injuries, such as disc herniation, disc degeneration, fracture of the vertebral body, tears in muscle ligaments, and strains and sprains. In many cases, these injuries lead to permanent, partial, or total disability.

The estimated direct costs of disabling back injuries and low back pain to US industries is approximately $20 billion per year. The indirect costs may be up to four times this amount. These costs quite likely are underestimated. Surgical procedures alone, for example, are estimated at $12 billion per year.[1] If that is true, the direct and indirect costs may be much higher than the current estimate of $100 billion per year.

Regardless of the amount, the costs are primarily shared by industries, insurance companies, and related gevernment agencies. Once these costs are incurred, the capital is no longer available for investment, economic growth, and prosperity. The outcome is a significant detriment to industrial productivity. Besides the economic burden, humane reasons mandate control of MMH-related overexertion injuries.

A number of alternatives have been suggested to control overexertion injuries caused by performing manual materials handling tasks, such as:

- Eliminate the need for strenuous MMH jobs
- Reduce or minimize MMH job demands
- Eliminate or minimize stressful body postures and/or movements
- Use mechanical equipment to move materials
- Train workers in proper material handling techniques
- Train workers to enhance their physical strength and endurance
- Design MMH jobs that can be performed by workers without undue risk of overexertion injuries.

Some of these alternatives, however, may not be feasible in any number of work settings. Lack of available space and investment capital, for example, would prevent mechanization. Under these circumstances, designing MMH jobs that can be performed by workers with minimal risk of injury may be the only practical solution.

MMH JOB DESIGN PROCEDURES

One of the popular approaches to designing MMH jobs is based on the capabilities of a certain population percentile. The majority of design procedures developed to date, however, are for a specific material handling activity such as lifting, pushing, and carrying. In reality, the majority of industrial jobs are multiple activity tasks and involve two or more different manual material handling activities. In such cases, most existing procedures are inadequate.

A comprehensive procedure for designing or evaluating new or existing multiple activity MMH jobs can reduce risk potential of stressful elements to 1.0 or less.[2,3] The procedure involves dividing the job or task into basic MMH activity elements, such as lifting, lowering, pushing, and carrying, and then analyzing each individual element individually, for risk potential. Elements with a risk potential greater than 1.0 are considered stressfull and must be redesigned. The formula to determine risk potential is:

$$\text{Risk Potential} = \frac{\text{Actual (Required) Work Rate}}{\text{Acceptable (Desired) Work Rate}}$$

The procedure utilizes data generated in several previous studies[4-6] and is constrained by several conditions:

- Workers use free-style body posture rather than a restricted posture.
- Normal workday is no longer than 12 hours with 30-minute rest breaks after 4 hours (lunch) and 10 hours (supper) and 10-minute rest breaks after 2 hours, 6 hours, and 8 hours. Workers are also free to pause for routine necessities, such as take a drink of water or go to rest room.
- The MMH capabilities of the working population are normally distributed.

The Design Procedure

The following illustrates this approach, using a step-by step design procedure:

Step 1. Break the job (task) into basic manual materials handling activity elements: lifting, lowering, pushing, pulling, and carrying. Note: There may be more than one element per activity.

Step 2. For each element, from Table 3-1 and Figure 3-1, determine the input variables and their values. For job evaluation, the values for the existing job should be used. For new jobs, these values should be based on anticipated working population gender, dimensions of the container/package or box to be handled, and some preliminary information on the workplace layout.

Step 3. Based on the values of input variables (Step 2), determine, from Table 3-2, the acceptable capacity for each element for 50th percentile population. These capacities are for box size and distance (vertical and horizontal) conditions shown in Table 3-2.

Step 4. Adjust the acceptable capacities determined in Step 3 for the effect of box size and distance by multiplying them with appropriate percentages given in Tables 3-3, 3-4, and 3-5. If the box has no handles, reduce the acceptable capacities by 7.2%. If body turning/twisting is involved (up to 90°), reduce the acceptable capacities by 8.5 %. If boxes are to be placed accurately on a narrow shelf or if the box is to be guided slowly to its final destination, reduce the acceptable capacities further by 10%.

Step 5. Adjust the Step 4 acceptable capacities for the population percentile desired by using the standard deviation models and the standard normal deviates given in Tables 3-6 and 3-7, respectively.

Step 6. Determine acceptable work rates (kg-m/min) by multiplying Step 5 acceptable capacities by handling frequency and travel distance.

Step 7. Adjust Step 6 acceptable work rates for the effect of working duration by reducing them by percentages given by the following relationships:

% for Males = 101.42 - 3.4 x Work Duration (in hours)

% for Females = 100.81 - 1.94 x Work Duration (in hours)

Step 8a. If designing a new job, STOP.

Step 8b. If evaluating an existing job, determine actual work rate for each element:

Actual Work Rate = Actual Weight x Handling Frequency per Minute x Distance Travelled

Step 9. Calculate risk potential, R , for each element by dividing the actual (required) work rate with the acceptable (desired) work rate.

Table 3-1. Input Variables and Their Units.*

Activity	Input Variable	Unit
All	Sex Box length (L) Frequency (F)	Male or female Centimeters No. of handling/minute
Lifting	Height level (H) Vertical distance (V)	=1, Floor-knuckle (LFK) =2, Knuckle-shoulder (LKS) =3, Shoulder-reach (LSR) Centimeters
Lowering	Height level (HL) Vertical distance (VL)	=1, Knuckle-floor (LOWKF) =2, Shoulder-knuckle (LOWSK) =3, Reach-shoulder (LOWRS) Centimeters
Pushing	Horizontal distance (HP)	Meters
Pulling	Vertical height (VH)	Centimeters
Carrying	Horizontal distance (HC)	Meters

* Also see Figure 3-1 for definitions of height level, vertical distance, and box length. Range for H and HL are: 0 cm to 84 cm for LFK and LOWKF; 84 cm to 135 cm for LKS and LOWSK; and 135 cm and above for LSR and LOWRS.

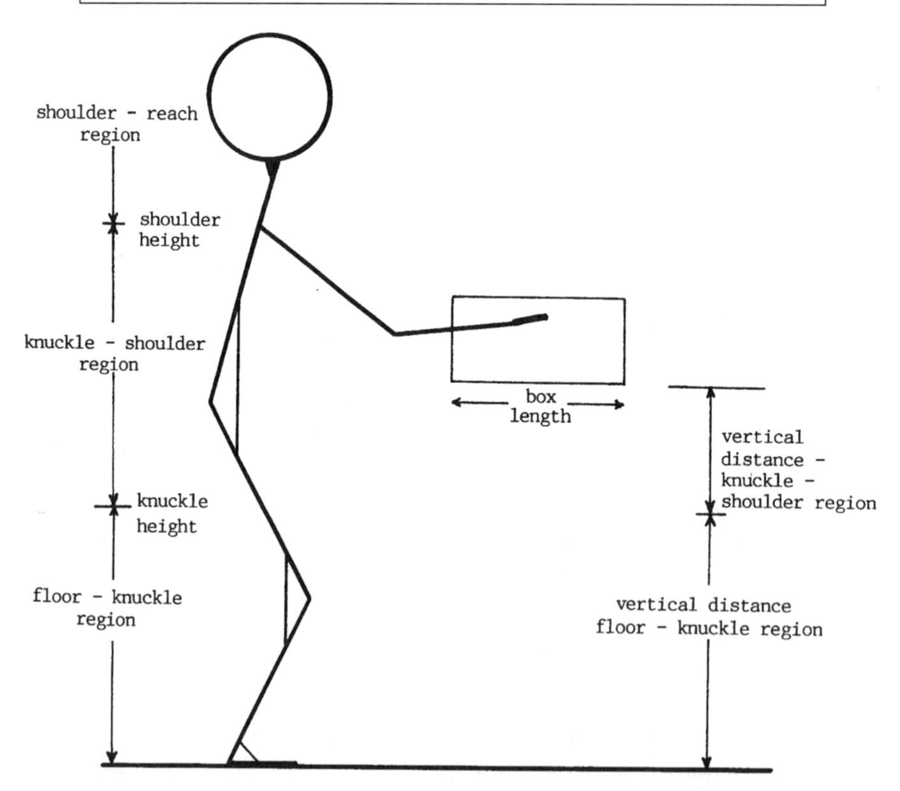

Figure 3-1. Definitions of Height Level, Vertical Distance, and Box Length.
Note: Floor-knuckle region is approximately 82-83 cm. For pulling activities,vertical height is the height of hand from the floor.

Table 3-2. Acceptable Capacity (Kg) Models for 50th Percentile Population.

Activity	Sex	Model	Conditions*
LFK	Male	27.94EXP (–.052) (F)	L=49 cm V=76 cm
	Female	15.96EXP (–.040) (F)	Same as above
LKS	Male	27.51EXP (–.043) (F)	L=49 cm V=51 cm
	Female	14.36EXP (–.027) (F)	Same as above
LSR	Male	22.17EXP (–.047) (F)	Same as above
	Female	13.06EXP (–.054) (F)	Same as above
LOWKF	Male	32.42EXP (–.060) (F)	Same as above
	Female	19.64EXP (–.054) (F)	Same as above
LOWSK	Male	24.25EXP (–.027) (F)	Same as above
	Female	14.71EXP (–.005) (F)	Same as above
LOWRS	Male	20.28EXP (–.039) (F)	Same as above
	Female	13.05EXP (–.014) (F)	Same as above
Push	Male	29.32EXP (–.055) (F)	HP=2.1 m
	Female	21.08EXP (–.067) (F)	Same as above
Pull	Male	28.65EXP (–.056) (F)	VH=0.6 m
	Female	21.19EXP (–.032) (F)	Same as above
Carry	Male	39.86EXP (–.056) (F)	HC=2 m
	Female	22.86EXP (–.036) (F)	Same as above

*These conditions are exact. The models accommodate a large range of frequency (F). Range for F is: 1 to 12 for lifting and lowering; 0.1 to 10 for carrying; 0.1 to 10 for pushing/pulling for males and 0.1 to 5 for pushing/pulling for females. Out-of range frequencies, such as 15 lifts per minute, may be used provided they are within 25% of the bounds.

Table 3-3. Box Size Adjustment (%) for Lifting and Lowering Activities.*

	Box Size from 36 cm to 49 -cm		Box Size from 49 ±cm to 76 cm	
	Male	Female	Male	Female
LFK	+.70	+.01	–.58	–.71
LKS	+.58	+.54	–.01	–.01
LSR	+.22	+.60	–.02	–.01
LOWKF	+.79	+.93	–.52	–.52
LOWSK	+.60	+.41	–.01	–.01
LOWRS	+.53	+.44	–.01	–.01

Note: + means that step 3 capacity should be increased by this percentage; – means that step 3 capacity should be reduced by this percentage.
* The unit of multipliers given in the table is percent per centimeter. In order to determine the exact adjustment for a box, table values should be multiplied by the difference in actual box length and 49 cm and then divided by 100.

Table 3-4. Vertical Distance Adjustment (%) for Lifting and Lowering Activities*.

	Distance Between 25 cm and 51 -cm		Distance between 51± and 76 cm	
	Male	Female	Male	Female
LFK	+.38	+.74	+.14	0
LKS	+.73	+.88	–.49	–.30
LSR	+.58	+.72	–.34	–.33
LOWKF	+.65	+.53	–.08	–.19
LOWSK	+.75	+.75	–.28	–.34
LOWSR	+.67	+.74	–.29	–.39

Note: + means stat step 3 capacity should be increased by this percentage; – means that step 3 capacity should be reduced by this percentage.
* The unit of multipliers given in the table is percent per centimeter. In order to determine the exact adjustment for actual vertical distance, table values should be multiplied by the difference in actual vertical distance and 51 cm and then divided by 100.

Step 10. Evaluate Rs as follows:
- If R< 1 for all elements, the job (task) is not expected to result in any overexertion injury and is acceptable.
- If R> 1 for any element, the job (task) must be redesigned by reducing the actual work rate. Actual work rate may be reduced by reducing the weight or force being handled, reducing the handling frequency, and reducing the travel distance.

Table 3-5. Distance Adjustment (%) for Pushing, Pulling, and Carrying Activities.

Activity	Sex	Adjustment Percentage
Pushing	Male	88.66 EXP(–.011) (HP)
	Female	90.91EXP(–.010) (HP)
Pulling	Both	131.49EXP(–.004) (VH)
Carrying	Male	100.27EXP(–.018) (HC)
	Female	99.70EXP(–.012) (HC)

Range for HP is from 2.1 m to 60 m. For HC, the range is from 2 m to 8.5 m. Range for VH is from 60 cm to 150 cm.

Table 3-6. Standard Deviation (Kg) Models for Various MMH Activities.*

	Male	Female
LFK	6.52–.30F	3.48–.12F
LKS	6.75–.26F	2.52–.07F
LSR	5.56–.18F	1.58–.02F
LOWKF	9.60–.43F	3.22–.08F
LOWSK	7.22–.16F	3.12
LOWRS	6.04–.19F	1.88
Pushing	12.57–.56F	6.63–.84F
Pulling	8.27–.36F	5.72–.06F
Carrying	12.71–.56F	4–.04F

*Step 4 capacities are for the 50th percentile population. In order to determine capacities for other population, assume that MMH capacities are normally distrinbuted. The capacity for any population percentile, thus, can be determined from the relationship: Capacity for a given population percentile = 50th percentile capacity + normal standard deviate for the population percentile (given in Table 3-7) x standard deviation (given in this table).

A Case Study

The stacking and palletizing job in a cattle feed bagging plant, located in the midwestern United States, has been analyzed. The job involves stacking feed bags (22.73 kg each) on a wooden pallet (20.32 cm high) located approximately 1.38 m from the conveyor (Figure 3-2). A male operator lifts the bag off the conveyor, carries it to the pallet, stacks it, and then returns for the next bag. Forty-five bags (five per layer, total nine layers) are stacked per pallet. Bags are 15.24 cm thick, 40.64 cm long (dimension in the sagittal plane - dimension away from the body), and 76.20 cm wide (dimension in the frontal plane which divides the body into two halves, front and back). While carrying, the effective bag dimension in the sagittal plane is reduced, due to deformation, to 38.10 cm. Bags are carried at a height of 81.28 cm. It takes 575 seconds to completely load the pallet (handling frequency of 4.69 bags per minute), and the worker operates 450 minutes each day.

Complaints of persistent arm, shoulder, and back pain warranted evaluation of the job. The step-by-step procedure, described above, was used. The goal was that the job should be acceptable to the upper 75% of the population. The job was broken into 11 MMH elements and for each element actual and acceptable work rates and the risk potential were calculated.

Based on the pallet loading time, approximately 153 minutes are spent in grabbing and lifting the bags off the conveyor, 153 minutes are spent in carrying them to the conveyor, and lifting them (top five layers) takes another 90 minutes. Lowering the bags for the bottom four layers takes about 54 minutes. Duration for each layer was determined by dividing the total activity time in proportion to the vertical distance travelled. Table 3-8 shows the input values, actual and acceptable work rates, and risk potential for all 11 elements.

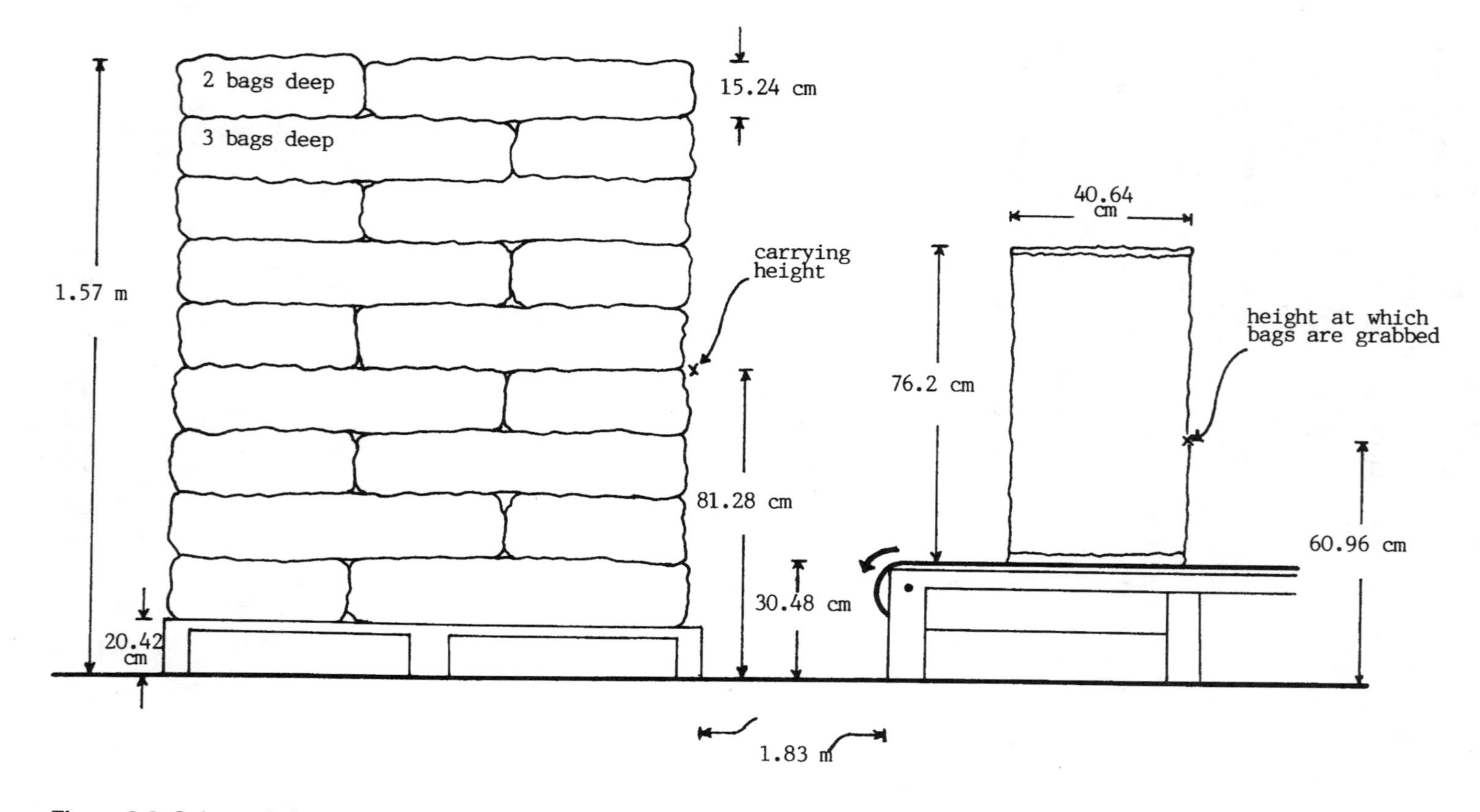

Figure 3-2. Schematic Layout of the Workplace.

Table 3-7. Standard Normal Deviates for Various Population Percentiles.*

Population Percentile	Standard Normal Deviate
95	–1.645
90	–1.282
85	–1.036
75	–0.674
50	0
25	+0.674
15	+1.036
10	+1.282
5	+1.645

*Reproduced from a normal distribution table.

Calculation of acceptable work rate and risk potential for Elements 1 and 2 follow:

Element 1
Acceptable Capacity = 27.94 EXP(-.052) (4.69) From Table 3-2
x [1+.70(49-38-1)÷100] From Table 3-3
x [1+.38(76-20.32)÷100] From Table 3-4
-.6745(6.52-.30 x 4.69) From Tables 3-6 and 3-7
= 26.28 kg
Acceptable Work Rate = 26.28 x 4.69 x .2032
= 25.045 kg-m/min
Adjusting this for work duration gives the acceptable work rate
=25.045[101.42-3.4(153/60)]÷100
=23.23 kg-m/min Given in Table 3-8
Element 2
Acceptable Capacity = 39.86 EXP(-.056) (4.69) From Table 3-2
x 1 (since HC is less than 2 m)
- .6745(12.71-.56 x 4.69) From Tables 3-6 and 3-7
= 25.22 kg
Acceptable Work Rate = 25.22 x 4.69 x 1.83
= 216.45 kg-m/min
Adjusting this for work duration gives the acceptable work rate
= 216.45[101.42-3.4(153/60)]÷100
= 200.76 kg-m/min Given in Table 3-8

Table 3-8. Job Analysis for the Case Study.

Element	Input Data	Acceptable Work-rate (kg–m/min)	Actual Work-rate (kg-m/min)	Risk Potential
1. Lifting from conveyer	L=38.1 cm V=20.32 cm F=4.69/min H=1 Duration=153 min	23.23	21.66	0.93
2. Carrying from conveyor to pallet	HC=1.83 m F=4.69/min Duration=153 min	200.76	195.06	0.97
3. 1st layer lower	L=38.1 cm VL= 53.35 cm F=4.69/min HL=1 Duration=23 min	56.82	56.85	1.00
4. 2nd layer lower	L=38.1 cm VL=38.1 cm F=4.69/min HL=1 Duration=16 min	44.92	40.61	0.90
5. 3rd layer lower	L=38.1 cm VL=22.86 cm F=4.69/min HL=1 Duration=10 min	30.08	24.36	0.81
6. 4th layer lower	L=38.1 cm VL=7.62 cm F=4.69/min HL=1 Duration=5 min	11.05	8.12	0.73
7. 5th layer lift	L=38.1 cm V=7.62 cm F=4.69/min H=1 Duration=5 min	8.90	8.12	0.91
8. 6th layer lift	L=38.1 cm V=22.86 cm F=4.69/min H=2 Duration=18 min	28.16	24.36	0.86

Table 3-8 (part 2). Job Analysis for the Case Study.

Element	Input Data	Acceptable Work-rate (kg–m/min)	Actual Work-rate (kg-m/min)	Risk Potential
9. 7th layer lift	L=38.1 cm V=38.1 cm F=4.69/min H=2 Duration=18 min	41.84	40.61	0.97
10. 8th layer lift	L=38.1 cm V=53.35 cm F=4.69/min H=2 Duration=25 min	51.81	56.85	1.10
11. 9th layer lift	L=38.1 cm V=68.58 cm F=4.69/min H=3 Duration=30 min	46.76	73.10	1.56

As shown in Table 3-8, the risk potential of Elements 10 and 11 exceeded 1.0. These Elements, therefore, were considered excessively stressful. The least expensive solution in this case was to eliminate the top two layers of bags from each loaded pallet. This reduced the number of pain related complaints significantly—from frequent to occasional.

References

1. Helms CA. CT of the spine: an overview. In *Low Back Pain: Solving the Clinical Challenge.* Secaucus, NJ: The Network for Medical Education Publication; 1985;467:5-12.

2. Mital A. Generalized model structure for evaluating/designing manual materials handling jobs. *International Journal of Production Research.* 1983;21:401-412.

3. Jiang BC and Mital A. A procedure for designing/evaluating manual materials handling tasks. *International Journal of Production Research.* 1986;24:913-925.

4. Mital A and Asfour SS. Material handling capacity of workers. *Material Flow.* 1983;1:89-99.

5. Mital A and Fard HF. Psychophysical and physiological responses to lifting symmetrical and asymmetrical loads symmetrically and asymmetrically. *Ergonomics.* 1986;29:1263-1272.

6. Mital A and Wang LW. Effects on load handling of restricted and unrestricted shelf opening clearances. *Ergonomics*; 1989; 32.

Anil Mital is an Associate Professor of Industrial Engineering and Founding Director of the Ergonomics Research Laboratory at the University of Cincinnati. He holds a BE degree in Mechanical Engineering from Allahabad University, India and MS and PhD degrees in Industrial Engineering from Kansas State University and Texas Tech University, respectively. Dr. Mital is a senior member of the Institute of Industrial Engineers and a former Director of the Institute's Ergonomics Division. His professional interests are in the areas of applied ergonomics, application of systems methodologies to ergonomics, metal cutting, sensor design, and traditional industrial engineering.

4

Ergonomics in the Management of Occupational Injuries

Tarek M. Khalil, PhD, PE; Elsayed Abdel-Moty, PhD; Shihab S. Asfour, PhD; Renee Steele-Rosomoff, RN, MBA; and Hubert L. Rosomoff, MD, D Med Sc

LOW BACK PAIN

Backaches can strike people of all classes, sexes, ages, races, education status, and professions. Low back pain (LBP) ranks first among all health problems in frequency of occurrence. An estimated 70% to 80% of the people in industrialized countries will develop some form of back pain during their lifetime.[1] Annual estimates of the new LBP cases range from 10% to 15% of the population.[2]

Back impairments are the most frequent cause of activity limitation in people under the age of 45, and they rank as the third most common cause of disability, after heart and arthritic conditions, in people aged 45 years and older. In the United States back pain is second only to headache as the symptom that prompts visits to physicians.[3]

Low back pain is often associated with functional disability, economic and social consequences, and enormous burdensome effects upon industry, health care systems, and society. Its costs, in lost earnings, worker compensation, and disability payments and medical care expenses exceed those for any other single health disorder. In 1978 an estimated $14 billion were spent for the treatment of LBP in the United States.[4]

A person with LBP can expect about $40,000 in medical expenses and loss or interrupted income and related benefits.[5] While these costs represent direct losses, many safety researchers and loss prevention specialists recognize that indirect losses, such as lost time, replacement of employees and equipment, and lower profit and productivity, are up to four times the direct costs.

The impact of LBP is expected to hit even harder in the next decades as longevity and the average age of the work force increase. Lost productivity, humanitarian considerations of the injured individual, the billions of dollars spent each year on treatment and surgeries, and the issue of vocational restoration are but some of the staggering aspects that make the prevention of low back pain and its treatment both a priority and a challenge to industrial and health care professionals.

In their effort to resolve the LBP problem health care providers have advocated a myriad of medical treatment approaches. The success of agressive noninvasive treatment of chronic LBP in restoring function and reducing pain is well recognized. This treatment approach requires a multidisciplinary team of health care professionals, since no one physician has the expertise or resources to manage this complex condition. The rehabilitation team usually includes physicians, physical therapists, occupational therapists, nurses, vocational counselors, psychologists, and psychiatrists. Ergonomics is equally important in successful LBP rehabilitation and management efforts.[6] The University of Miami Comprehensive Pain and Rehabilitation Center (CPRC) is an ideal example of the comprehensive team approach to resolving low back pain at the physical, psychological, and occupational levels (Figure 4-1).

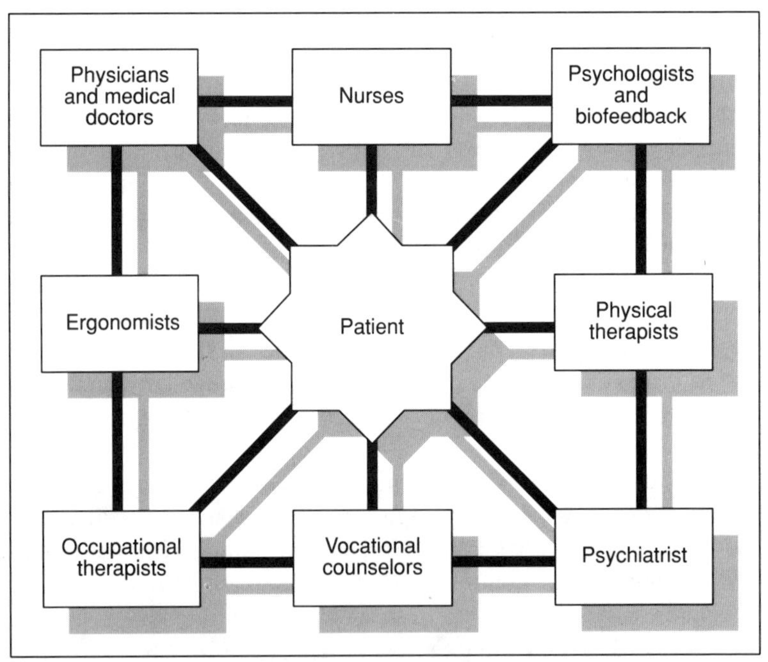

Figure 4-1. At the University of Miami CPRC, the Patient is the Focus of an Interdisciplinary Integrated Team Approach.

ERGONOMICS IN LOW BACK PAIN MANAGEMENT

Ergonomics addresses the LBP problem at three critical stages: preinjury, postinjury rehabilitation, and postrehabilitation. The preinjury stage employs ergonomics guidelines as the best approach to avoiding injury, by reducing musculoskeletal stresses and preventing accidents.

During the postinjury rehabilitation stage, the most effective approach, both medically and economically, is the quick restoration of the functional abilities. Re-engineering of the work environment and specific tasks during the postrehabilitation stage, permits a safer, more compatible relationship between people with LBP and their working or living environment.

Ergonomic contributions at the CPRC encompass many areas (Figure 4-2), including:

- Determining the LBP patient's status through a performance profile of functional abilities
- Comparing the patient's functional abilities to the physical demands of the job
- Developing techniques to restore functional abilities through improved performance of the musculoskeletal system, such as muscle re-education and functional electrical stimulation
- Developing patient education programs based on proper body mechanics according to biomechanical models and ergonomic concepts
- Assisting in the selection of jobs that match the measured functional capabilities of rehabilitated people
- Evaluating the effectiveness of treatment regimens on the restoration of functional abilities and pain reduction
- Evaluating industrial work sites to deter injury
- Designing or recommending appropriate seating devices, equipment, and furniture to reduce work-related stresses
- Developing realistic job simulations during rehabilitation to permit patients to simulate tasks under medical supervision

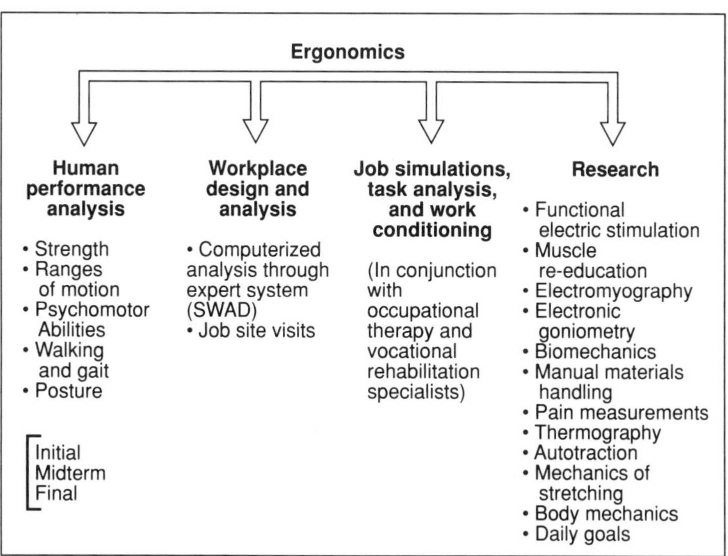

Figure 4-2. A Chart Showing the Various Activities Within the Ergonomics and Bioengineering Division at the CPRC.

EXAMPLES OF ERGONOMICS APPLICATIONS IN REHABILITATION

The following case studies, drawn from actual CPRC applications, demonstrate ergonomic contributions to the management of low back pain patients. Over the past decade, the CPRC has evaluated or treated approximately 6,000 LBP patients.[7] Treatment at the CPRC consists of the following:

- An active, aggressive, and intense physical medicine program
- Patient education
- Drug-intake reduction or elimination
- Physical and occupational conditioning
- Dedicated behavioral medicine
- Vocational and recreational rehabilitation with a therapeutic goal of an immediate return to a productive lifestyle
- Ergonomics consultation for human performance analysis, job simulations, workplace design, and functional electrical stimulation

Case Study One
Using the Human Performance Profile To Quantify Rehabilitation Effectiveness

A female electrician, aged 22, installs and repairs electric fixture wiring, apparatus, and control equipment. She cuts and bends threads, assembles and installs electric conduit using hand and power tools, and also tests circuits and observes equipment functioning. While carrying a box of tools she tripped over a cable, landed on her tailbone (coccyx), and felt immediate pain. She suffered lower, as well as upper back pain, and was admitted to the CPRC for treatment.

Initial ergonomic evaluation showed a significant loss in her functional abilities due to pain and disability (Table 4-1). Her trunk flexion was limited and static strength values were much lower than those of healthy women in the same age group. Her walking pace was slower than normal, and she had limited tolerances for sitting, standing, and walking.

After four weeks of aggressive therapy, final ergonomic evaluation showed a significant increase in all measures of her functional performance. The most dramatic result was an over 100% increase in the upper extremity and back strengths. She was able to flex her trunk and touch her toes with no difficulty. She was also able to walk faster, and to squat fully, and her posture and hand steadiness improved.

The patient's ability to perform activities of daily living reflected these changes. Her tolerances returned to normal, she was able to sit for more than one hour, stand for one hour, and walk one mile in 30 minutes. She was discharged from the CPRC with a pain level of 1/10, (10% of initial pain level), and returned immediately to full-time employment with no restrictions.

The human performance profile (HPP) helped by documenting her condition on admission to the rehabilitation program. As a result, therapists could determine areas of weakness requiring special attention, follow-up on her progress throughout her clinical stay, and document the success of her rehabilitation.

Table 4-1. Human Performance Profile Showing the Changes in Measures of Functional Performance for Case Study #1.

Measure	Initial	Final	Change
1. Grip Strength, lb	60	70	+17%
2. AME, lb.			
Composite	90	120	+33%
Leg	160	216	+35%
Shoulder	35	68	+94%
Arm	26	55	+112%
Back	42	127	+202%
3. Trunk Flexion, deg.	70	120	+71%
4. Walking Speed, ft/sec	5.9	6.5	+11%
5. Functional Tolerances:			
Sitting, min.	25	60	+140%
Standing, min.	25	60	+140%
Ambulation, ft/min	167	176	+5.4%
6. Reaction Time, sec:			
Simple	0.29	0.23	+26%
Choice	0.70	0.52	+35%
7. Hand Steadiness, #	10	4	+60%
8. Squatting Ability, (inches from floor)	15	0	+100%
9. Posture Score (out of 100)	85	95	+12%
10. Reported Pain Level (Scale 1 to 10)	7	1	

Case Study Two
Ergonomic Analysis of a Sitting Workplace

A woman, aged 32, weighing 120 lb, and measuring almost 5 ft 3 in tall, is the director of operations for an office supplies company. She had suffered chronic back discomfort since 1982 with two previous surgeries. She did well for five years but has had recurring, increased low back pain with leg radiation on the left side. She became less active, had difficulties wearing high heels, and used medications without benefit. She also had a high level of chronic tension. She was admitted to the CPRC for evaluation and treatment.

On admission, the ergonomist analyzed the tasks and simulated her job environment in order to recommend modifications of work conditions. She was asked to provide dimensions and photographs of her office and a description of her

continued next page

workday. Pacing, work-rest schedule, prolonged sitting in meetings and at her desk, and poor body mechanics during work were identified as major problems in task performance. Problems in the physical environment (Figure 4-3a) were related to the location and angle of the VDT screen and the keyboard, the location of the telephone, and chair design. In particular, her chair was not compatible with her anthropometric dimensions, especially seat depth, elbow rest height, seat height, and seat width.

Via the use of electromyogram (EMG) biofeedback and task and workplace simulation, the ergonomists determined that doing the job in the current workplace arrangement caused stress on the muscles of the neck and lower back (Figure 4-4). The task performance was adjusted and modified in the simulation, and the woman was instructed on the use of proper body mechanics for reaching, answering the telephone, using the computer terminal, opening and closing drawers, and handling files.

The work surface layout was also modified to adhere to principles of motion economy, optimal reaches, and proper body mechanics (Figure 4-3b). The computer terminal was moved to the left side and because of space limitations, the keyboard was placed in a slide-in holder. The telephone was moved to the left-frontal side within comfortable reach. Eventual use of a head set was advised in order to free her hands and to avoid poor habits associated with prolonged telephone use. Because it was used infrequently, the modem was placed to the left of the computer terminal. All other work tools (pens, stapler, calendar, tape, adding machine) were relocated within easy reach.

In addition, a computer analysis, using the expert system Sitting Workplace Analysis and Design (SWAD), was performed to obtain optimal workplace parameters and dimensions for the woman's anthropometric dimensions.[8] This analysis provided a comparison between the workplace before and after modifications and the potential stresses associated with using the current facility (Table 4-2). The woman's case related directly to the stress points indicated by SWAD (Figure 4-5). Based on the analysis, an adjustable foot rest and a new chair were recommended.

The patient moved steadily through the rehabilitation program and worked diligently to regain her function. She was continuously exposed to task simulation activities and by the end of three weeks of rehabilitation her overall functional status had improved significantly. She returned to work immediately following discharge from the CPRC without any restrictions. Her work setting was modified as recommended, and she was fully satisfied with her new environment. A follow-up showed that she continued to enjoy her work with no major difficulties.

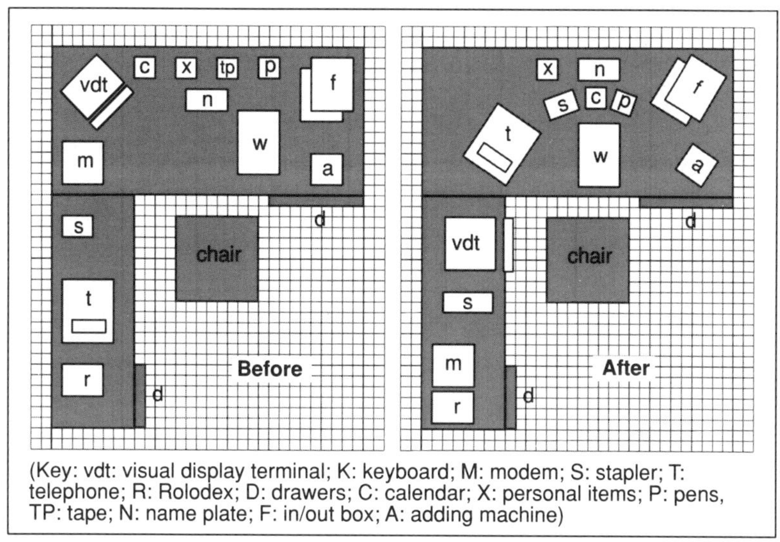

Figure 4-3. Layout of the Work Surface of Case #2 Before and After Ergonomics Analysis.

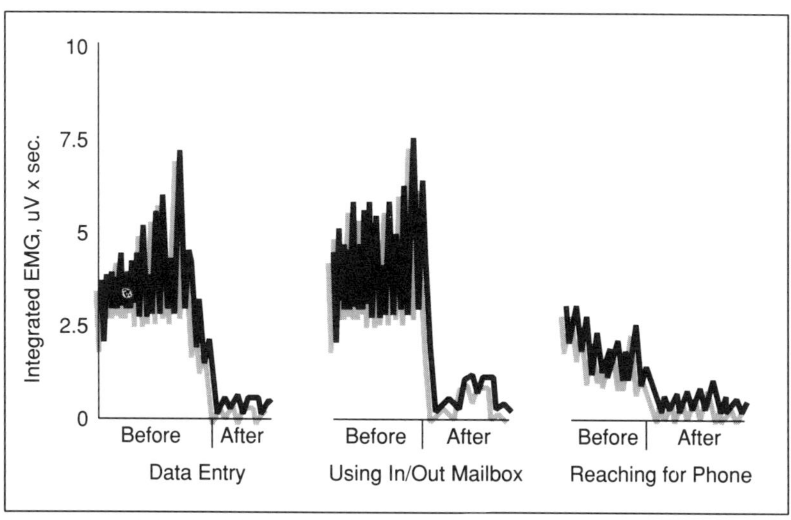

Figure 4-4. Effect of Changing Some of the Workplace Parameters on the Electromyographic Activity (EMG) of the Trapezius Muscles of the Neck While Case #2 Was Performing Simulated Tasks.

Table 4-2. SWAD Provided a Comparison Between the Dimensions of the Workplace Before (Actual) and After (Recommended) Modifications.

Workplace Design Features		
Item	Actual Dimension	Optimal Dimension
Chair		
Back rest length	20.00	25.00
Back rest width	21.00	> 8.59
Seat height	18.50	18.50
Seat depth	18.50	15.50
Seat breadth	21.00	20.00
Desk		
Height from floor	29.00	28.50
Width	66.00	> 43.49
Depth	30.00	> 21.74
Knee clearance	6.00	2.50
Foot rest height	0.00	2.50
Note: All dimensions are in inches. Use figure to identify parameters.		

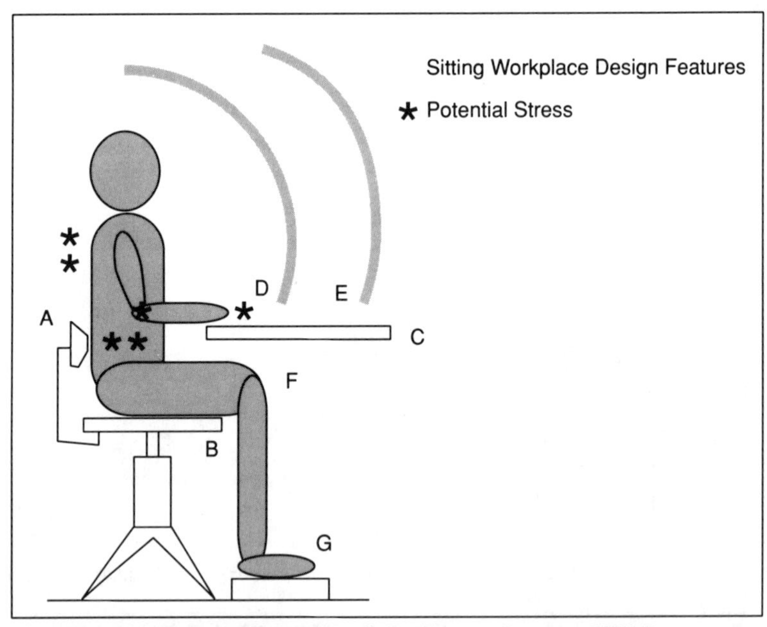

Figure 4-5. Computer Printout of the Expert System SWAD Indicating the Locations of Potential Stress Points of the Body Due to the Use of the Workplace of Case #2 Before the Application of Modifications/Adjustments.

Case Study Three
Restoration of Muscle Function

A male carpenter and metal worker, aged 51, standing 6 ft tall and weighing 230 lb, had a series of minor back complaints from 1969 to 1973. In 1975, he slipped and twisted his body and felt immediate low back pain. In 1976, he underwent a laminectomy and discectomy at L4-5. Although he felt better for only four to five months following his surgeries, he was never pain-free. Medications, bed rest, and occasional physical therapy continued intermittently until 1979.

Since then he was without directed treatment, was taking medications, and was inactive. He was told that he needed a second back operation, but it was postponed. At the time he was admitted to the CPRC for rehabilitation in 1986, he was in constant pain, which he described as severe and burning. He could only stand or sit for 10 minutes, and could walk only one-eighth of a mile.

The initial multidisciplinary evaluation showed a slow, maladaptive gait with difficulty in heel and toe walking. His coordination and equilibrium were good, and his motor system was intact except for what appeared to be residual weakness in the right extensor hallucis longus and dorsiflexors group. All muscle activity in the right lower extremity was tremulous and his reflexes were depressed at the ankle on the right side. There was no evidence of active motor nerve root compression. He then entered the full rehabilitation program.

As a part of his physical conditioning and strengthening, he received functional electric stimulation to restore function to the weak right-ankle dorsiflexion muscles. Functional electric stimulation (FES) is a technique in which an external electrical stimulus is used to induce functional tetanic contraction of a muscle or muscle group. This technique has been extensively investigated and has proved to be very useful in the restoration of muscle strength and working capacity.[9]

This technique was applied to the right tibialis anterior muscle five times a week for two weeks. Each FES session consisted of 15 induced isometric tetanic muscular contractions (attempted ankle dorsi flexions) against maximum resistance. Each contraction lasted for 15 seconds (on time) followed by a rest period of 35 seconds. The stimulus was delivered in the form of rectangular biphasic electric current of 20 Hz frequency or pulse rate; 225 microseconds pulse width or duration; three seconds rise time; and tolerable amplitude (intensity). Tolerance was recorded and strength and EMG data were recorded to document neuromuscular condition and progress.[10]

Results of evaluating the tibialis anterior muscles' function before and after the FES application showed a drastic loss in muscle function before treatment (Table 4-3); the strength of the right dorsiflexors was only 5 lb as compared to the strong performance of the left side (MVC = 30 lb). Electromygram activity of the right tibialis anterior muscle was very depressed, and EMG amplitude measures showed a 1:10 ratio between the recruitment level of the right and left tibialis anterior muscles. The power spectrum of the surface EMG also showed a significant reduction of the signal's energy contents at all frequency bands (Figure 4-6).

During the first application of FES, the patient tolerated 75 mA, which increased to 200 mA during the final session. After the initiation of FES treatment, there was a gradual improvement in the patient's ability to voluntarily dorsiflex and

continued next page

evert his right ankle. After two weeks of administring FES, strength and EMG values had increased dramatically (Table 4-3); isometric muscle strength of the weak right dorsiflexors approached that of the stronger, healthier left side. The EMG of the right tibialis anterior resembled that of a normal interference pattern (Figure 4-6).

Measures of the amplitude contents of the EMG signal during the final evaluation showed a more than four-fold increase in value when compared with those obtained during the initial evaluation. Also, there were changes in the measures of the frequency component of the myoelectric signals which indicated increased firing of the motor units. The patient finished his rehabilitation program successfully. One year later, he was still fully functional, having maintained all muscle gains achieved during the program.

Table 4-3. Summary of Measures of Functional Abilities for the Right and Left Tibilais Anterior Muscles Case #4.

	Right		Left	
Measure	Initial	Final	Initial	Final
1. Maximum voluntary contraction, lb	5	32	30	36
2. Calf girth, Cm	35	37.2	38	38
3. Tolerance, mA	75	200	N/A	N/A
3. Mean rectified, mV	5.9	33.1	51.7	53.4
4. Full wave rectified integral, mV. sec	23.0	128.4	200.3	214.7
5. Energy, mV*mV	26.9	735.2	821.7	856.0
6. Sum of voltage excursions, mV	23.7	132.5	206.6	261.3
7. Number of amplitudes	122	466	687	751
8. Number of turning points	1155	1414	1592	1426
9. Band width, Hz	102.8	118.3	134.35	128.0
10.Median frequency	62.5	58.6	31.5	35.2
11.Average firing frequency, Hz	91.4	113.3	113.8	113.1

DISCUSSION

The staggering statistics on the magnitude of the LBP problem make its prevention and treatment a compelling, humane, and economic priority. It is possible for people with even the most severe disability to return to full function with a properly designed multidisciplinary therapeutic program. The contributions to the management of LBP and the methods for doing so have been developed according to ergonomic fundamentals.

Quantitative measurement using the ergonomics approach are sensitive enough to detect small, yet significant clinical differences in functional abilities among patients or across time within a single patient. The usefulness of these evaluations in detecting changes in functional aspects of chronic LBP patients are a valuable addition to any comprehensive rehabilitation effort. Ergonomic measures should be used in conjunction with those performed by physicians and therapists in order to assist in diagnosing and designing suitable treatment programs.

The human performance profile determines the effectiveness of treatment and physical readiness to return to gainful employment and other activities of daily living. At the University of Miami CPRC, ergonomists work daily in the hospital environment with patients as well as other members of the health care team to restore function to people with low back pain. The ergonomics team members participate in many phases of the patient rehabilitation program, such as job simulations, education of patients regarding proper body mechanics, and in advancing rehabilitation research. The capacity to utilize ergonomics principles in a clinical setting leads to the optimal treatment and improvements in the long-term prognosis of people with low back pain.

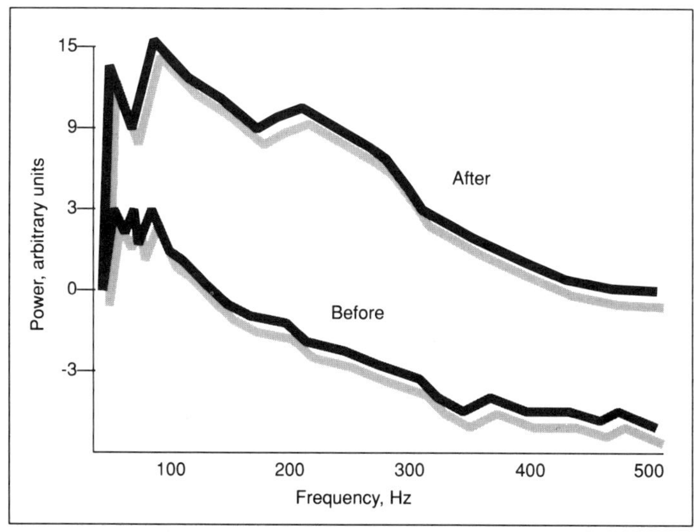

Figure 4-6. Power (frequency) Spectrums of the EMG of the Right Tibialis Anterior Muscle of Case #3 Before and After the Administration of FES.

References

1. Locke TC. Stretching away from back pain injury. *Occupational Health and Safety.* 1983;7:8-14.

2. Steinberg GG. Epidemiology of low back pain. In: Stanton-Hicks M. ed. *Chronic Low Back Pain.* New York, NY: Raven Press; 1982;1-13.

3. Cypress BK. Characteristics of physician visits for back symptoms: a national perspective. *American Journal of Public Health.* 1983;73:389-395.

4. Schaepe JL. Low back pain: an occupational prespective. In: Stanton-Hicks M, ed. *Chronic Low Back Pain.* New York, NY: Raven Press; 1982.

5. Nordby E. Epidemiology and diagnosis in low back injury. *Occupational Health and Safety.* 1981;1:38-41.

6. Khalil TM, Asfour SS, and Moty EA. New horizons for ergonomics research in low back pain. In: Eberts RE and Eberts CG, eds. *Trends in Ergonomics/Human Factors.* North Holland/Elsevier Science Publishers BV; 1985;591-598.

7. Rosomoff HL, Green C, Silbert M, and Steele R. Pain and low back rehabilitation program at the University of Miami School of Medicine. In: Lorenzo KY, ed. *New Approaches to Treatment of Chronic Pain: A Review of Multidisciplinary Pain Clinics and Pain Centers.* NIDA Research Monograph 36, Department of HHS; 1981.

8. Abdel-Moty E, and Khalil TM. Computer aided design of the sitting workplace. *Computers and Industrial Engineering.* 1986;11(suppl.):1(4):23-26.

9. Abdel-Moty E, Khalil TM, Rosomoff HL, and Steele R. Functional electrical stimulation for low back pain patients. *Pain Management.* 1988; November-December:258-263.

10. Abdel-Moty E, Khalil TM, Rosomoff HL, and Steele R. Computerized signal processing techniques for the quantification of muscular activity. *Computers and Industrial Engineering.* 1987;11(3):193-203.

Tarek M. Khalil is professor and chairman of the Department of Industrial Engineering at the University of Miami. He is also chief of the Ergonomics Division of the University of Miami Comprehensive Pain and Rehabilitation Center and is the Director of the Environmental Health and Safety Program of the University. Dr. Khalil holds a BME degree from the University of Cairo and his MSIE and PhD degrees were earned from Texas Tech University. Dr. Khalil is a Fellow of the Institute of Industrial Engineers and served on its Board of Trustees from 1982 to 1984. He is also a member of the Human Factors Society, ASSE and AIHA. He is the author of more than 150 publications and his research work has been funded by NSF, the US Army, Aeromedical Research Laboratory, NIOSH, AID, and other federal, state and private organizations.

Elsayed Abdel-Moty is a Senior Research Associate in the Department of Neurological Surgery at the University of Miami School of Medicine. He received a BS degree in Electronics and Computer Engineering and a MS degree in Biomedical Engineering from Cairo University. His PhD in Biomedical Engineering was earned from the University of Miami-Coral Gables. His research interests include electromyography, factors related to muscle fatigue, biomechanical analysis, thermography, digital signal processing, low back pain, functional capacity evaluation, expert systems and computerized workplace design, and functional electrical stimulation. Dr. Abdel-Moty is a member of the Institute of Industrial Engineers.

Shihab S. Asfour is Professor of Industrial Engineering and Neurological Surgery at the University of Miami. He received his PhD in Industrial Engineering frm Texas Tech University. Dr. Asfour is a senior member of the Institute of Industrial Engineers and a former president of the IIE Miami Chapter. Asfour is also a member of the Ergonomics Research Society, the Human Factors Society and the Society of Manufacturing Engineers. Dr. Asfour's research interests include industrial ergonomicrk physiology, biomechanics, rehabilitation engineering, safety engineering, work measurement, and systems design.

Renee Steele-Rosomoff holds the following degrees and certifications: Bachelor of Science in Nursing from Florida International University; Masters in Business Administration from Barry University; Certified Rehabilitation Counselor; Certified Insurance Rehabilitation Specialist; and Certified Rehabilitation Registered Nurse. Rosomoff is also an adjunct assistant professor in the Department of Neurological Surgery, Schools of Medicine and Nursing at the University of Miami. Rosomoff is also a Past President of the National Association of Rehabilitation Nurses and has authored over 35 publications including, *You've Had a Heart Attack, So What!*

Hubert L. Rosomoff is the founder of investigation of laser on the brain and the first to introduce the use of lasers for treatment of tumors in man and was the first to perform brain surgery with the use of laser in 1965. Rosomoff received his MD from Hahnemann Medical College and his D Med SC in Physiology from Columbia University. He also holds Diplomate status with the National Board of Medical Examiners and the American Board of Neurological Surgery.

5

Bagging Dry Chemicals: A Case Study

John L. Wick

Bagging dry chemicals has recently become automated, but powders with certain properties defy current automatic bagging systems. The problem of achieving an accurate fill weight with powders that have inconsistent flow properties requires that the bagging operation be semiautomatic.

In this case study, one person operated two semiautomatic filling machines that could fill a 50 lb bag with a plus or minus 10% accuracy. The worker removed a filled bag from one machine, carried it to a check-weigh scale, placed an empty bag on the machine spout, adjusted the weight of the filled bag, and carried it to a powered take-away conveyor. This was done during the fill cycle but required slightly more time than the other machines. The second machine was tended during the fill-cycle of the first machine. The two lines were mirror images and converged on the take-away conveyor. The operation ran day and evening shifts; cleaning and maintenance were done on the third shift.

The sequence of operation of the two machines follows:

1. Remove bag from machine - line two
2. Place new bag on machine - line two
3. Check/adjust weight - line two
4. Close bag spout - line two
5. Carry to conveyer - line two
6. Remove bag from machine - line one
7. Place new bag on machine - line one
8. Check/adjust weight - line one
9. Close bag spout - line one
10. Carry to conveyer - line one

During a period of less than two years, seven workers were removed from the job because of carpal tunnel syndrome, a type of wrist injury and a common cumulative trauma disorder. Various job rotation schemes were tried with no effect. An in-depth ergonomic study was then commissioned to determine the causes of the injuries and to develop solutions.

The task analysis methodology [1] included:

- Videotaping the job. The worker was videotaped from both sides, front and back while performing the job.
- Performing a detailed task analysis. While viewing the videotape in slow-motion, the job was divided into elements, and the positions of the hands, arms, shoulders, and trunk were recorded for each element.
- Collecting workplace data. The ergonomists measured the vertical and horizontal layout of the work area. They asked the worker to demonstrate the forces on the finger and hand dynamometers to determine the finger or hand forces needed to handle the bags.
- Evaluating the data. The ergonomists used the results to determine damaging wrist motions (DWM = the wrist not straight with a pinch or power grip); excessive pinch grip forces (50% of maximum ability[2]) and fifth percentile's maximum ability[3]; compressive force on L5/S1 disc[4]; and metabolic energy expenditure.[5]
- Identifying the ergonomic problems. Those job elements that exceeded the maximum criteria were singled out for change.
- Designing solutions. The ergonomists developed an appropriate workplace design to solve as many of the problems as possible.
- Implementing changes. The new design was fabricated, installed, and fine-tuned as necessary.
- Validating the new design. Data were taken from videotape and workplace measurements of the new design to ensure the changes solved the problems and did not introduce new ones.

DETAILED DESCRIPTION OF THE JOB BEFORE CHANGE

Before the job was redesigned, productivity was 600 bags per shift. The workers agreed this was the maximum that could be done without excessive fatigue. The cycle time of the filling machine ranged from 18 to 27 seconds depending on the product's viscosity. The machine cycle began when the operator pushed a start button. The bag rested on a saddle that sensed the weight. The machine stopped filling the bag when the desired weight was reached.

The operator placed the receiver spout of an empty bag on the filler spout of the machine. The filler spout was 40.5 in from the floor. The operator had to abduct and flex both shoulders and flex and ulnar deviate both wrists with pinch grips to place the bag on the spout.

The filled bag was removed from the machine by grasping the top fold with the left hand and wrapping the right arm around the bag. The operator had to abduct and flex both shoulders, extend and ulnar deviate the left wrist with pinch grip, and flex and ulnar deviate the right wrist with pinch grip. The task required 12 lb of finger force.

The bag was carried 30 in to the platform check-weigh scale. When resting on the scale, the bottom was 14.5 in from the floor and the top was 24 in from the scale platform. The carry was done with the same hold used to remove the bag from the filler and imposed the same shoulder and wrist angles and finger forces.

The standard weight was 50 lb plus or minus 1% and were 50% inaccurate. An operator could adjust weight by either removing or adding the product using a narrow cast aluminum scoop through the bag spout, which had a 3 in diameter. The use of the scoop required abduction of the right shoulder and extreme flexion and ulnar deviation of the right wrist. The scoop was held in a power grip with 57 lb of finger or hand force. An average of two times per inaccurately filled bag (overall average of one time per bag) was required to achieve accuracy.

The operator then closed the bag by folding the spout into the top (flex and abduct the right shoulder and flex the right wrist). The bag was then carried to the take-away conveyer. The operator grasped the bag with both hands in pinch grips at the top fold, lifted it off the scale, walked 28 in, and lifted it to the surface of the conveyer 20 in (50.8 cm) from the floor. The carry required extreme flexion and extreme ulnar deviation of both wrists and pinch forces of 12 lb.

IDENTIFIED PROBLEMS AND CAUSES

As shown in Table 5-1, most of the criteria used for design of jobs and workplaces were exceeded. The damaging wrist motions were caused by the filler spout and the bag on the scale being too high and the weight of the bag pulling the hand down during the carries.

Table 5-1. Summary of Task Analysis Data: Maximum vs. Actual.*

	Maximum	Actual
DWM	1,000/hr	1457/hr
Pinch grip force	8 lb	12 lb
Power grip force	25 lb	57 lb
L5/S1 Disc compressive force	515 lb	787 lb
Energy expenditure	3.0 kCal/min	3.1 kCal/min

* Summary of totals required for one cycle.

High pinch grip forces were caused by loose-fitting gloves, powder-coated surfaces, and heavy loads handled with pinch grips. The high power grip force required to use the scoop was due to a high friction surface of the cast aluminum and blunt edge. The excessive compressive force on the L5/S1 disc was caused by the weight of the load and the forward trunk angle when the bag was lifted from the scale. The weight of the load and the distances carried were the primary causes of energy expenditure.

DESCRIPTION AND RATIONALE FOR CHANGES

Productivity could possibly be improved if energy expenditure could be lowered, since a time study revealed the machines were idle 75% of the job cycle. A platform covering the entire work area was installed to effectively lower the spouts of the filling machines. The platform was 14.75 in high. The scales were placed on the platform on a shortened stand and the take-away conveyor was raised. Roller conveyors were installed from the filler saddle to the scale, on the scale, and from the scale to the take-away conveyor. All heights were calculated to keep the top of the bag at elbow height for the 50th percentile female.[6]

The roller conveyor eliminated having to carry the bag, with the associated high compressive forces on the L5/S1 disc, extreme wrist angles, and high pinch grip forces. It also minimized the energy required to perform the job, because the operator could push the bag along the conveyor from the filler to the scale to the take-away conveyor.

The elbow height of the 50th percentile female was chosen because it was not practical or cost-effective to make the system adjustable and most operators were female. The height chosen allowed an operator to perform the tasks of placing the empty bag, removing the filled bag, adjusting the weight, and moving the bag on the roller conveyor with neutral shoulder and wrist positions.

THE REDESIGNED JOB

Although the job sequence did not change, productivity was increased to 800 bags per shift. The operator placed an empty bag on the 36 in high filling machine spout, which required flexion of the left wrist and extension of the right wrist. The filled bag was slid off the spout and saddle onto the roller conveyor and pushed to the platform check-weigh scale. The task could be done with straight wrists and low finger forces. The weight adjustment required the same high power grip forces, but was performed with neutral shoulder and wrist positions. The operator pushed the bag from the scale on the roller conveyor to the take-away conveyor and tipped the bag onto the conveyor belt, all with straight wrists.

The redesigned job resulted in:

- Elimination of shoulder flexion and abduction
- Reduction of damaging wrist movements for each hand from 1457 per hour

to 114 per hour
- Reduction of pinch grip force from 12 lb to 4 lb
- Reduction of disc compressive force from 787 lb to 101 lb
- Reduction of energy expenditure (at new production rate) to 2.8 KCals/minute
- Increase in production from 600 bags/shift to 800 bags/shift (33.3%)

An unexpected positive result was two employees who were off the job with carpal tunnel syndrome were able to permanently return. The operators also reported less fatigue after the change at 800 bags per shift than before the change at 600 bags per shift.

Two types of problems remained. The problem of grip force was unresolved by the new design because an alternate, practical method or tool for adding or removing product from the bag could not be developed. A problem resulting from the new design was reports of back discomfort. The fixed height of the points of operation resulted in imperfect fits for many operators—an expected outcome when a workplace is designed for the average person.

DISCUSSION

The practical application of ergonomics is effective in preventing injuries and improving productivity. An ergonomics practitioner, however, must ensure a practical approach and an effective method of implementation. Such guidelines include effective problem-solving techniques; for example, thoroughly analyzing the job to properly define the problems, carefully identifying the causes, developing practical solutions, and validating the solutions to ensure they are appropriate. Practical solutions are those which solve the problems without creating new ones, are cost-effective, and can be implemented in a reasonable amount of time without difficulty. This project was not unique in its success; the methodology has been used many times with similar results.

References

1. Drury CG and Wick JL. Ergonomic applications in the shoe industry. *Proceedings of the 1984 International Conference on Occupational Ergonomics*. Toronto, Canada: 1984;489-490.

2. Grandjean E. *Fitting the Task to the Man*. London: Taylor and Francis Ltd.; 1982;29.

3. Van Cott HP and Kinkade RG, eds. *Human Engineering Guide to Equipment Design*. Washington, DC: American Institute for Research; 1982;565.

4. Fish DR. *Practical Measurement of Human Postures and Forces in Lifting, Safety In Manual Materials Handling*. National Institute for Occupational Safety and Health; 1972;72-77.

5. University of Michigan, 1988.

6. Diffrient N, et al. *Human Scale 1/2/3*. Cambridge, MA: The MIT Press; 1981.

John L. Wick is affiliated with J&J Consulting in Seattle, WA. Previously Mr. Wick was Senior Staff Ergonomist for Chesebrough-Ponds, Inc. He developed a corporate-wide ergonomics program, training more than 400 managers, engineers and operators resulting in $7 million in savings in five years.

His extensive experience in ergonomics covers work measurement systems, manual and semi-automated assembly, materials handling processes, heat stress management, sewing, and office design.

Initially trained in Business Management, he received his training in Ergonomics under internationally known Ergonomist Dr. C. G. Drury, Professor at State University of New York at Buffalo. Mr. Wick has published in numerous articles and is a member of the Human Factors Society.

6

Screening and Training As Administrative Controls

T. L. Doolittle, PhD
Karl Kaiyala, MSPE

Ergonomics has two primary goals: to improve worker well-being and to improve productivity. The results of meeting these goals is to reduce the number of mishaps that can injure workers or damage products, equipment, or facilities. The most common approach to achieving these goals is to fit the job to the worker emphasizing job design or redesign in the case of existing jobs, which is compatible with known human morphologic, physiologic, and psychological norms.

Controls ergonomists use to fit the job to the worker fall into two broad categories: engineering controls and administrative controls. Engineering controls are the usual method of choice. Examples of engineering controls, which ensure the compatibility of on-the-job lifting requirements with human capacity for safe lifting, include providing mechanical or power assists, partial, or complete automation, and ensuring that the objects to be lifted weigh less than an amount determined to be acceptable to the majority of the population, for any specific lift.

For existing jobs, implementating engineering controls is not always feasible. As an alternate, ergonomists use administrative controls. These normally are accomplished through adopting policies and procedures, such as establishing a mass limit that workers are permitted to lift by themselves, and providing and requiring the use of another worker to assist in lifts that exceed the established limit.

Some tasks do not lend themselves to cooperative lifting. In cases in which these controls are not practical, job and worker matching is an administrative alternative. When the job cannot be fit to the majority of workers, the use of a selection system that identifies and assigns individuals with sufficient capacity to safely perform is a valid ergonomic procedure. Specialized training to increase the capacity of individuals, so as to increase the potential pool from which qualified workers may be selected, is an additional administrative control that complements screening. This case study describes the development of such a screening system and a complementary training program.

BACKGROUND

Ergonomic job analyses were done to determine the strength and endurance capacities required to effectively perform the job of a metropolitan firefighter. Predicated upon these requirements, a test battery was designed, screening standards were developed, and a training program was instituted to improve the strength and endurance of individuals so they could attain the standards. While this case study describes an experience with firefighting, and specifically affirmative action procedures to qualify women as firefighters, the methodology also has been applied to numerous occupations that have arduous manual material handling tasks.

The mayor of a western US city issued an edict that the impending recruit class for the fire department could not begin unless it included women. None of the 65 women who took the entry test could pass the physical agility component. The response was to substantially lower the passing levels for two of the physical agility test items. The time allowed to complete the hose-carry-tower-run event was increased by 240%, and the length of rope that had to be pulled in one minute at a tension of 50% was reduced by 18%.

As a result of this step, eight women were included on the civil service eligibility register and six were admitted to the recruit class. None completed the class: five resigned in lieu of dismissal and the sixth was dismissed. A lack of physical strength was cited as a reason for each termination. While this is a documented experience of one city, it was by no means unique or isolated. Numerous organizations across the nation, in an effort to meet affirmative action goals, have lowered strength and endurance standards.

The original test battery had been used for a number of years. Although it had never been validated formally, it had been effective in screening out individuals without sufficient strength and endurance to successfully complete firefighter recruit training. The lowering of standards negated the effectiveness (empirical validity) of the tests. One common complaint of the women, who were unaware that the standards had been lowered was, "We passed your test so we should have been able to make it through recruit training." While they tended to blame the training as being too hard, in reality the test standards were too low.

The mayor continued his policy and dictated that the next recruit class could not begin without female representation. It was clear from the prior experience, and knowledge of efforts made by other fire departments, that simply lowering standards to admit women to recruit training would not be a satisfactory solution. Although it would meet the mayor's edict and women would be in the class, the probability that they would complete the training was virtually nil unless appropriate actions were taken. The fire department's training chief recognized that supplemental training was a potential solution.

FEASIBILITY STUDY

Three of the women from the original class applied for reinstatement on the civil service register. Before granting the request, the feasibility of their becoming successful firefighters was evaluated.

Ergonomic Job Analysis

Prior to evaluating the womens' potential, the physical requirements of firefighting were analyzed to quantify the job requirements. This analysis began with a review of the hose and ladder sections of the fire department's *Manual of Fire Suppression*, describing and illustrating the required movements and tasks. The ergonomists observed firefighters performing selected physically demanding movements. They also identified each task's movement pattern and major contributing muscle group and then verified them through direct observation. Forces required to lift, carry, and manipulate hoses, ladders, and other equipment were obtained. These procedures identified the strength capacities that individuals must have to be able to function as firefighters.

The requirements were described in terms of common weight training movements; for example, floor-to-knuckle (midthigh height when standing erect) lifting is necessary in moving ladders, preparing to carry a stretcher, picking up hose, and other tasks. These movements employ essentially the same muscular actions that are used in the squat exercise. Likewise, a number of tasks require lifting from knuckles to shoulders and from shoulders to reach (arms overhead). The upright row and biceps curl exercises are representative of knuckle-to-shoulder lifts, and the military press is a shoulder-to-reach lift.

A compilation of routine firefighting tasks and the corresponding movements along with the required forces are shown in Table 6-1. Although not exhaustive, the list represents a variety of physically demanding activities that are typical of the job. All of the tasks require the exertion of an absolute amount of force that is dependent upon the mass of the object being lifted. The required force is independent of the individual's stature, age, gender, or race.

In addition to the exertion of absolute force, individuals must have capacity to carry and climb under laden conditions. Two distinct additional traits are involved. The first, *relative strength*, is the ability of individuals to lift their own body mass and is important in climbing and survival activities. The second, *cardiorespiratory capacity* or *aerobic power*, is the capacity to take in oxygen and produce metabolic energy. This capacity is typically expressed in milliliters of oxygen per kilogram of body weight per minute (mL/kg/min). Various investigators[1,2,3] have suggested that for individuals to be successful firefighters they should possess a maximum oxygen consumption capacity (aerobic power) in relation to their body mass well above that of the average adult.[1-3] Values of 42 to 45 mL/kg/min, with higher capacities being desirable, have been recommended as minimums.

Table 6-1. Firefighting Tasks, Associated Movements and Masses.*

Firefighting Tasks	Exercise			Movements†			
	MP	BC	Sq	DL	LP	UR	FR
Raising ladder	34	34	34	A			
Hanging smoke ejector	34		34	A		34	P
Lifting stretcher		P	38	A			
Shoulder loading hose & standing		30	30				
Hoisting equipment from above		P				44	
Hoisting equipment from below					37		
Pompiering roof ladder	P	P					12
Extending fly on extension ladder					37		
Lateral ladder shift			61				
Replacing equipment on truck	34	34					

* Value in the table are kilograms that typically have to be lifted, carried or pulled; P indicates partial range of motion, and A indicates the lift is an alternative to the squat.

† Exercise movements and associated ergonomic terminology:

MP= Military press (shoulder to reach)
BC = Biceps curl (knuckle to shoulder with elbow flexion only)
Sq = Squat (floor to knuckle squat lift)
DL = Dead lift (floor to knuckle stoop
LP = Lap pull (reach to shoulder--pulling down)
UR = Upright row (knuckle to shoulder with shoulder abduction)
FR = Forward rise (primary action is shoulder flexion)

Evaluating the Women

Each woman was evaluated for maximum strength in each movement, and her performance was compared to the levels deemed to be neccessary for success. The maximum strength measures were determined with standard barbells and repetitive trials until a mass was obtained that could be lifted through the full range of motion once but could not be immediately repeated. This amount is operationally termed *one repetition maximum* (1-RM), and generally is accepted to be the best measure of an individual's maximum strength for the specific lift. In order to perform a task satisfactorily and safely, an individual's capacity should exceed the force required by the task by one-third. Table 6-2 lists these corresponding capacities.

Table 6-2. Strength Requirements of Firefighters.*

Task	Exercise	Terminology	Job	Test
Lateral ladder shift	Squat	Floor to knuckle	61	82
Hanging smoke ejector	Military press	Shoulders to reach	34	45
Hoisting from above	Upright row	Knuckles to shoulder**	44	59
Lifting	Biceps curls	Knuckles to chest†	34	45
Sum				231

*Masses (kg) that have to be lifted on the job and corresponding one repetition maximum test standard (see text);
**With shoulder abduction;
†Elbow flexion only.

This reserve capacity is necessary because of three factors. First, the strength tests are conducted under ideal circumstances in terms of body position, footing, distribution of mass, and ease of grasping the object. On the job, any or all of these circumstances may be changed, which effectively increases the force that must be exerted. In short, on-the-job lifting can never be any easier than the test condition, and it may be far more difficult. Second, the job requires that any given task be followed by another. There is, therefore, little or no time to recover from a truly maximum effort.

Third, people are more prone to injury the nearer they are to this lifting capacity. Cady et al[4] and Doolittle and Kaiyala[5] have shown that weaker firefighters are more prone to on-the-job musculoskeletal injuries. Chaffin et al,[6] while not concerned with firefighters, stated, "Weaker workers when performing high strength requiring activities, have an increased incidence and severity of musculoskeletal and contact injuries".

Subsequently, Chaffin et al[7] reported that the incidence and severity of injuries increase at a ratio of about 3:1 as the job strength requirements approach the worker's capacity. Ayoub et al[8] indicated that 75% of capacity is the critical level above which the rate of injury raises dramatically. If the task requirement is not to exceed 75% of an individual's capacity, then the capacity must be at least 1.33 times the requirement.

TRIAL PROGRAM

Knowing the difference between each woman's capacity and that required for all of the lifts resulted in a conclusion that an extensive strength training program could enable them to succeed in recruit training. The ergonomists estimated that the training program would require 20 weeks for two of the women and up to 35 weeks for the third to build sufficient strength. These estimates were based on expected increases of 2% per week usually observed in an intense strength training program. Indications were that one woman, despite gaining sufficient strength in 20 weeks, probably would be at a mechanical disadvantage because of her short stature. The third candidate would also have to lose weight to be successful.

The ergonomists further pointed out that the women would need considerable self-motivation and dedication for the training program, which required heavy weight training sessions of 60 to 90 minutes three to four times per week and 30-minute running sessions, to develop aerobic power, two to three times per week. Little, if anything, could be done to make the training fun, and while expected gains were reasonable there could be no guarantee that they would occur.

In addition to the issue of strength, the fire department's training chief was mindful of the mechanical, manipulative problems experienced by the women during training. He clearly saw a need to train the women in mechanical and manipulative skills before they could attend recruit school. He proposed that the department install a women's prerecruit training program whose content and duration would produce meaningful improvements in physical fitness and mechanical, manipulative ability. This meant a full-time paid program. A prerecruit program was started with the three women. They understood that they would have to attain the requisite levels of muscular strength (Table 6-2) and pass the Civil Service Firefighter Test, without adjustment, to be admitted to the recruit class that was scheduled approximately seven months later.

An individualized training program was prescribed for each woman. Detailed written instructions and verbal explanations, regarding the amount of mass and the number of times it was to be lifted were provided for each exercise. Similar instructions were given for running and stretching routines. Monday-Wednesday-Friday routines consisted of 11 weight training exercises and one bout of running. Tuesday-Thursday regimens included three bouts of running and five additional strength exercises. Each woman maintained a daily record that detailed what she had done on each exercise. Guidelines were provided so that they knew when and how to increase the intensity for each exercise. They were evaluated monthly and their individual exercise prescriptions modified accordingly.

After six weeks one women resigned. Her separation came as little surprise, because she shunned the strength training and was unable, or unwilling, to adhere to the weight reduction plan recommended for her. Three weeks later the shorter of the remaining two women sustained a back injury and was eliminated from the program.

The third woman persevered. She received intensive one-on-one firefighting instruction and continued participation in the prescribed physical conditioning program. She had to work out alone from that point on, except for the monthly evaluation sessions to modify her exercise prescription. At one point, she altered her training regimen and consequently, there was no improvement between evaluations. This was pointed out to her as well as the importance of adhering to the exercise prescription.

After five months, she entered a second prerecruit training program consisting of herself and six minority men. This was a six-week, full-time firefighting training program, but it did not include formal strength or aerobic training. She was encouraged to continue strength training and running, on her own. Subsequently, she entered recruit class for a second time and as result of the intensive training program, she completed the class and was assigned to a firefighting company.

DEVELOPMENT OF FORMAL PROGRAM

The success of the trial program with one of the three women eventually led to a formalized prerecruit training program, but not without difficulties. A sizable recruitment effort was undertaken and 249 women attended orientation classes; 99 of these took an entry test and 28 passed it at a level that the Civil Service Commission had set specifically for the prerecruit program. The Commission had ignored the ergonomically derived requirements and projections for improvement in setting its prerecruit standards. In order to be admitted into recruit school, however, the women would later have to pass the physical agility test (Civil Service Firefighter Test) at the minimum standards. This was designated to be a promotion examination.

The ergonomists evaluated 12 women after they had been admitted to the first formal prerecruit program, which was intended to prepare its members for a recruit class scheduled to begin in five weeks. The assessment of the women's strength levels in relationship to the ergonomically determined required lifting capacities showed that such a goal was totally unrealistic. The initial evaluations indicated that with a five-month training program six of the individuals had reasonable potential for attaining sufficient levels of fitness, while the remaining six were estimated to have very low probabilities for success. A minimum of five months was needed for training if any of the women were to have a reasonable chance of attaining sufficient strength.

Political pressure to get women into the fire department was intense at this time, which is why those estimated as having very low probability of success were retained in the program. It was also the reason that only five weeks were originally allocated to the prerecruit training program, since that is when the next class was scheduled to begin and there had not been adequate preplanning.

Arguments ensued, but reason did prevail and the city relented. The 11 women (one had quit after three weeks) in the five-week program were given the option of entering into the then-starting recruit class or continuing the prerecruit class for an additional five months. All of them opted for the latter and two more women joined the program. The ergonomists noted at the time that future employment decisions should be predicated on the ergonomic standards, since only about one-half of the 13 women otherwise hired had any reasonable chance of success.

Prerecruit Class

The prerecruit training was conducted five days per week. It consisted of approximately four hours per day of instruction in firefighting skills and basic mechanics and approximately three hours per day of strength and endurance conditioning. About one hour was devoted to commuting by bus between the two training sites. The women were divided into two groups to ensure adequate individual attention and access to the equipment. One group received technical instruction in the morning and fitness training during the afternoon, with the other group following the reverse format. Specific strength and endurance regimens, as described for the pilot program, were prescribed for each woman based on her capacities.

To prevent individuals altering their programs as they saw fit, a trained exercise specialist supervised the physical conditioning program. Each woman maintained a daily exercise record by recording the amount of weight and the number of repetitions for all assigned lifts and the distances and times for all aerobic activities. The records were periodically reviewed and the women were retested five times over the 29 weeks of the program. Individual exercise prescriptions were revised as necessary. The records and the retesting also served as motivation; each woman continuously knew where she stood in relation to the goals and how much progress she had made.

All of the women increased their strength and aerobic capacities significantly. A composite strength index was obtained by totalling the maximum amount that an individual could lift in the squat, upright row, biceps curl, and military press exercises (SUM). Table 6-2 shows that this SUM should equal 231 kg or more; none of the women attained that SUM. While they had made significant gains, they had not improved enough to meet the ergonomically determined standards.

Recruit Training

Because of political pressures 9 of the 11 women entered the recruit class, although they had not attained the strength goals. None of them satisfactorily completed it. Three, however, eventually became firefighters: one by completing a subsequent recruit class with the department after improving her fitness, and two in a smaller department that is slightly less physically demanding. One additional woman, the strongest (sum = 225 kg, only a 6 kg decrement), probably would have completed the class had she not voluntarily withdrawn because of a fear of heights. The fact that none of the women successfully completed the recruit class after failing to attain the requisite strength levels reinforced the credibility of the ergonomically determined requirements.

PREDICTIVE INDICES

The prerecruit program lasted six weeks longer than the five-months estimate. The projections for improvement were overly optimistic, although they were highly pessimistic compared with the Civil Service Commission's test results and the city's desires. It became obvious that realistic prerecruit training standards had to be determined for prerecruit training that were compatible with training duration and standards for the job of firefighter.

The original projections for improvement were predicated on an average of 2% per week, an amount cited in the literature as occurring with men over a 12-week program. Similar projections for women or for longer programs were not available. The women did average 1.8% per week for the first 14 weeks; however, after that the mean rate of increase began to plateau. For the four exercises that the SUM comprises the mean gain over the 29 week program was 1.4% week.

Assuming an increase of 2% per week, the prerecruit standards for the four SUM lifts should have been 152 kg if a person were to achieve 231 kg after 26 weeks. With only a 1.4% per week improvement, the starting level would have to be raised to 170 kg for a program of the same duration. The mean for the 11 women at the start of the program was 145 kg; and only six had SUMs of 150 kg or greater and only two exceeded the 170 kg level. The variability in percentage of improvement among the women and the significant plateauing led to reanalyzing the impact of lean body mass in an effort to improve the predictions for future groups.

A person's maximum potential for lifting heavy objects is a function of that person's lean mass. Lay people can interpret this by visualizing that it is more difficult to lift 75% of their own body weight than 50% of it. A person weighing 90 kg (198 lb) has an easier time lifting 45 kg (99 lb) than an individual weighing 60 kg (133 lb); this is the rationale for having weight classifications in Olympic weight lifting.

Since the adipose tissue contributes nothing to the individual's lifting capacity, it is more correct to relate strength to lean body mass. If the SUM is divided by the individual's lean body mass a ratio indicative of the person's relative strength is obtained. The mean ratio for the women at the start was 2.91 and had increased to 3.78 by the end of the program. In contrast, the mean ratio for 88 male firefighters selected at random was 3.60. Relatively speaking, the women were quite strong by the end of the prerecruit program, but their insufficient absolute capacity primarily was a function of their smaller size rather than their gender.

The mean increase in the ratio of SUM to lean body mass was 0.87, and ranged from 0.46 to 1.37. The lean body mass of the women increased (mean = 4.9%) as a result of the weight training program and explained the change in the ratio. The woman with the greatest absolute strength (SUM = 225 kg) had a ratio at the end of the program of 3.84, while a smaller woman with a ratio of 4.27 (the highest) was 11 kg lower in absolute strength. The latter woman had the smallest increase in the ratio (0.46), which indicated that she was very close to reaching her maximum potential. An alternative explanation would be that she did not put forth as much effort; however, subjective evaluations by the supervisors revealed that she was one of the most dedicated and hardest working individuals in the strength training program.

Further analysis revealed that the absolute increase in the SUM scores averaged 53.4 kg. When normalized on a per kilogram of lean body mass basis this was 1.05 (range 0.64 to 1.30 with a standard deviation of 0.29). *Lean body mass* is the fat-free portion of total body mass estimated with skinfold calipers or hydrostatic weighing techniques. SUM is the total of the masses lifted with the biceps curls, military press, upright row, and squat exerciese. These data enabled development of predictive indices based upon lean body mass and expectations from the normal curve (Table 6-3).

Table 6-3. Predictive Indices.

Lean Body Mass	Initial "SUM" and Associated Probability		
65 kg	131	153	164
Probability	0.05	0.30	0.52
60 kg	138	158	168
Probability	0.04	0.28	0.50
55 kg	145	163	173
Probability	0.04	0.26	0.48
50 kg	152	168	177
Probability	0.03	0.25	0.46
45 kg	159	174	181
Probability	0.03	0.22	0.42

Given the initial SUM for an individual with the corresponding lean body mass the probability of that person attaining a SUM of 231 kg after five months of intensive strength training is on the line below the associated SUM. For example, if two individuals both have lean body masses of 65 kg, the one that can lift a SUM of 164 kg has a 52% probability of reaching the 231 kg goal, whereas one who can lift only 131 kg has only a 5% chance of attaining the goal. In contrast, the person of 45 kg lean body mass with a SUM of 174 kg has a 22% probability of reaching the goal. Probabilities are based upon the required increase in the ratio of lean body mass to SUM and the differences attained by the women in this study.

EVALUATING THE MODEL

Luckily, for evaluating the indices, a new prerecruit class was about to begin and 17 women were sent for evaluation. Using the predictive indices, the ergonomists projected that seven had a reasonable chance for success, three were marginal, and the remaining seven should not be hired. The program began with seven women. For reasons known only to the city, those selected were: four from the "reasonable", one from the "marginal," and two from the "do not hire" categories.

The prerecruit program for these women was virtually identical to that earlier one, except that it ended up lasting 32 weeks. Six of the seven women who completed this prerecruit class took the standard firefighter physical agility examination. The one opting not to take the test had been in the "reasonable" category and had achieved the highest SUM (225 kg).

The test was treated as a promotion examination with entrance into recruit training conditional on passing it. Four of the six barely passed and entered recruit training along with the one women from the previous class who had improved her fitness level by working as a roofer. Of the two failing the test, one had been in the "do not hire" category. The other woman in this category did not satisfactorily complete the recruit class, although she did eventually become a firefighter. Of the other four who entered recruit training, three satisfactorily completed it and were hired by the department. The fourth accepted a position in another department before recruit training was complete.

In summary, neither of the "do not hire" candidates gained enough strength during the prerecruit class to complete recruit training, although one eventually became a firefighter. Three of the four rated as "reasonable" and the "marginal" candidate barely made it through recruit training. None of them had attained the expected SUM of 231 kg, although the four marginally successful women had achieved at least 89% of that goal. Since the successful women had been only marginally satisfactory, the goal of 231 kg for the SUM was retained.

The plateau effect again was observed, with virtually no increases occurring after the week 20 of training. The ergonomists recommended that programs be limited to 20 weeks, since that was the point of diminishing returns. By program's end the women had a mean ratio of SUM to lean body mass of 3.89, again indicating that they were quite strong. But as with the previous group, they were low in lean body mass.

The indices had improved the predictive power of using strength and anthropometric measures for determining which individuals would improve sufficiently during a five-month training program to meet valid fire department entrance standards. The fire department's training chief was able to convince the Civil Service Commission that the indices should be employed in hiring women for subsequent prerecruit classes. The chief emphasized that placing undersized individuals with totally inadequate strength who had virtually no chance of success in such programs was a waste of everybody's time and the city's money.

Two additional paid women's prerecruit classes were conducted, and no woman was hired unless she had a reasonable probability (greater than or equal to about 30%) of success based upon her initial levels of strength and lean body mass (predictive indices). The exercise specialist was replaced with one more knowledgeable of strength training in order to improve the day-to-day operation. All of the women in these two groups successfully passed the regular firefighter agility test.

At this point two changes occurred. The biceps curls and military press tests were added to the firefighter agility test battery, and the prerecruit classes became voluntary for women. The fire department provided the services of the same strength training expert for five to six months prior to scheduled examinations, but no longer paid women to participate. Women who took this opportunity were counseled by the trainer on the basis of the indices and most with little or no chance of success took themselves out of the program.

Application of the indices and the prerecruit training program, while still enforcing ergonomically sound standards, have resulted in a high success rate for all women admitted to subsequent recruit classes. A total of 56 women completed recruit school and probation to become firefighters in this city. This represents approximately 6% of the department nationwide, less than 1% of the professional metropolitan firefighters are women.

This case study has demonstrated that it is feasible to determine valid standards based on job demands, adjust those standards downward in accordance with anticipated gains to be made during training, and train individuals meeting the lowered standards to achieve the levels required by the job.

References

1. Dotson C, Santa Maria D, Davis P and Schwartz R. *The Development of a Job-Related Physical Performance Examination for Firefighters.* Report to National Fire Prevention and Control Administration; 1977;78-008.

2. Doolittle T. *Validation of Physical Requirements for Firefighters.* Technical Report for Seattle Fire Department; 1979.

3. Davis P, Dotson C and Santa Maria D. Relationship between simulated firefighting tasks and physical performance measures. *Medical Science Sports Exercise.* 1982;14(1):65-71.

4. Cady L, Thomas P, and Karwasky R. Program for increasing health and physical fitness of firefighters. *Journal of Occupational Medicine.* 1985;27(2):110-114.

5. Doolittle T and Kaiyala K. Strength and musculo-skeletal injuries of firefighters. *Proceedings of the Human Factors Association of Canada.* 1986;19:49-52.

6. Chaffin D, Herrin G, Keyserling W and Foulke J. *Pre-employment Strength Testing.* Cincinnati, OH: National Institute for Occupational Safety and Health. 1977;77-163.

7. Chaffin D, Herrin, G and Keyserling W. Preemployment strength testing: an updated position. *Journal of Occupational Medicine.* 1978;20(6):403-408.

8. Ayoub M, Selan J, and Jiang B. *Mini-Guide for Lifting.* Lubbock, TX: Texas Tech University; 1984.

T. L. Doolittle is an associate professor at the University of Washington. His major area of expertise is industrial ergonomics with emphasis on occupational fitness as it impacts worker productivity, industrial injuries, and affirmative action concerns. Doolittle was educated and earned his PhD from the University of Southern California.

Karl Kaiyala is the physical fitness director for a major metropolitan fire department. His major duties include regular physical fitness testing of the firefighters, providing exercise prescription, and conducting the department's pre-employment training program for prospective women firefighters. Kaiyala earned his BS and MS degrees from the University of Washington.

7

Cost Reduction Through Modification of a Heavy Physical Task

Jean M. Dalton
Nancy Smitten

Human factors problems associated with heavy materials handling tasks can be cost-effectively solved using ergonomics principles. This chapter examines a case study of a small group of electricians at Dofasco's hot strip mill who each week moved a 2335 lb circuit breaker. This particular task had a high potential for injury, and after a number of people were injured, this problem became a high priority in the hot strip mill electrical department. A human factors assessment was initiated to investigate the possible causes of injury and to make appropriate ergonomic recommendations to reduce the risk of injury.

PROBLEM IDENTIFICATION, INJURY AND TASK ANALYSIS

The first step was to identify specific problems of moving the circuit breaker by reviewing accident/injury data and a detailed task analysis. The data showed that shoulder and back pain were the most common complaints, with four injuries specifically attributed to moving the breakers (Table 7-1).

The task of switching breakers was videotaped and broken down into two main subtasks: pushing the breaker into the cell and pulling the breaker from the cell. The pushing subtask, which required that the worker move the breaker up an incline, could be broken down into five task components (Table 7-2). The pulling subtask, which was done by one person, could be broken down into four task components (Table 7-3).

Table 7-1. Injury Analysis.

Worker	Body Part Affected	Medical Visits	Days Off	Days On Modified
1	Left shoulder	6	2	25
2	Lower right back	1	-	-
3	Upper right back	1	-	-
4	Upper right back and shoulder	4	34	2

Table 7-2. Pushing Subtask.

Task Component	Ergonomic Concern
1. Position/align breaker with cell	Uneven floor Bulky breaker decreases visibility Track not easily seen at cell Large force required Difficult to control movement
2. Push breaker up ramp	Large force required Awkward posture
3. Push breaker into cell	Large force required Awkward posture Repeated push/pull required when breaker gets stuck
4. Attach crank	Stooped posture
5. Crank breaker into cell	Sustained bending Forceful movements of arm and shoulder Repeated elbow, wrist, and shoulder flexion and extension

Table 7-3. Pulling Subtask.

Task Component	Ergonomic Concern
1. Attach crank	Stooped posture
2. Crank breader to remove	Sustained bending Forceful movement of arm and shoulder Repeated elbow, wrist, and shoulder flexion and extension
3. Grasp handles	Stooped posture to reach low-level handles
4. Pull breaker from cell	Large force required Awkward posture

Although aspects of switching breakers, such as cranking, pushing, pulling, and bending, were identified as physically stressful, two aspects of the task required considerable physical effort. These aspects included moving the breaker into or out of the cell and aligning the breaker successfully, which often required several attempts.

Moving the Breaker

In order to push the breaker, the worker had to stoop, with the hands placed low on the breaker, to ensure that the breaker cover would not fall off (Figure 7-1). A similar position was taken when pulling the breaker since the handles were attached to the lower front of the breaker. In addition, the electrician had to keep one foot on a release latch while moving the breaker. Moving the breaker into the cell required the worker to push the breaker up a ramp, generating high forces in the electrician's lower back and shoulders.

Figure 7-1. Pushing the Breaker.

Aligning the Breaker

Workers reported that pushing the breaker into the cell was more difficult than pulling the breaker from the cell. Alignment problems and the awkward posture required to push the breaker up the ramp contributed to the problem. Alignment problems were due primarily to the breaker's large size. It was both awkward to maneuver and obstructed the worker's view of the cell opening. Being able to see the cell opening was important, because the electrician had to line up the breaker with a small track set back within the cell. As a result, the breaker would often get stuck, requiring repeated tries to maneuver it into proper position. Repeatedly pushing the heavy breaker up the ramp resulted in excessive stress on the lower back and shoulder.

BIOMECHANICAL ANALYSIS

The investigators used computerized and objectively measured biomechanical analysis of the pushing and pulling movements to objectively assess the levels of stress in the back and shoulders.

Low Back Forces

Pushing. The force required to push the circuit breaker was measured at the hand-equipment interface using two Revere UMP1-2.5-A 2500 lb range load cells calibrated up to 1000 lb and a four-channel recorder. Each measured trial was videotaped. The results are displayed in Tables 7-4 and 7-5.

Table 7-4. Pushing Forces.[1]

	Exerted Force (lbs)		Total Force	Max. Push Force (lbs)*		
	Rt. Hand	Lt. Hand		10TH%	50TH%	90TH%
Initial Push	125	83	208			
Breaker Push	150	105	255	65	114	159
Maximum Effort	180	120	300			

*At a height 64 cm; freq. 1/30 min.

Table 7-5. Computer Analysis.

	Total Pushing Force (lbs)	Joint Shearing Force (N)	Compressive Force (N)
Initial Push	208	-324*	2782
Breaker Push	255	-374*	2758
Maximum Effort	300	-421*	2736

*Negative shearing forces represent the L4 vertebra shearing posteriorly on L5.

The pushing forces measured at the hand were consistently high at three points:

- Initial push - the force required to overcome friction and start the breaker moving

- Breaker push - the force required to push the breaker onto the tracks within the cell
- Maximum effort push - the force required to dislodge the breaker when it gets stuck in the cell and realign it

Forces measured at the hand-equipment interface were compared with established standards,[1] which describe the maximum force that can be safely exerted by the 10th, 50th, and 90th percentile populations at various heights. A comparision of measured results showed that for each task condition, the total measured force exceeded the standard set for all percentile groups.

Critical postures, which concurred with the measured peak forces, were selected from the videotape and were simulated in a laboratory setting in order to take a sagittal view photograph for digitization (Figure 7-2). Joint markers were used to landmark the metatarsal, ankle, knee, hip, lower back, ear canal, shoulder, elbow, and wrist joints. The investigators used the WATBAK program, developed at the University of Waterloo, to assess the pushing, pulling, and lifting tasks in symmetric or asymmetric postures.[2] Application of the program required measuring the force needed to move the load (breaker) and to digitize the critical postures.

Figure 7-2. Sagittal View of a Pushing Motion.

Using this program, the investigators were able to biomechanically model the compressive and shearing forces at the lower back (L 4-5), elbow, shoulder, hip, knee, and ankle joints. Joint shearing and compressive forces represent the forces acting directly through and at right angles to the structures in the lower back. The following is a summary of their findings.

The calculated compressive forces approached the NIOSH action limit value (3433N) but did not exceed it (Table 7-5). An examination of this finding showed that the WATBAK program may have underestimated the forces occurring during a dynamic task, because no allowance was made for the effects of segment velocity and acceleration. Thus, the calculated measures represented a conservative estimate of the forces generated during the circuit breaker pushing task. Since the task also requires dynamic forceful movements, these forces probably exceeded the NIOSH action limits, which put the individual at risk of injury.

Table 7-6. Pulling Forces.[1]

		Maximum Pulling Forces				
	Total Force (lbs)	10th%	50th% (lbs)	90th%	Joint Shear Force (N)	Compressive Force (N)
Pulling Breaker	194.5	74	105	137	583	3496

Pulling. The force required to pull the breaker out of the cell was measured using a Strainsert Universal FL1U-25G 2500 lb range load cell calibrated up to 1000 lb. Peak forces were measured while initiating movement of the breaker out of the cell (Table 7-6).

A comparison of the WATBAK calculated compressive force, with standards developed by NIOSH[3], showed forces in excess of the action limit of 3433 N for the task. The National Institute for Occupational Safety and Health (NIOSH) recommends that engineering controls (job redesign) or administrative controls (employee education) be sought where the action limit is exceeded.

Table 7-7. Shoulder Forces.

	Initial Push (Nm)	Breaker (Nm)	Max. Push (Nm)
Right Shoulder Movement	53.9	63.2	74.3
Left Shoulder Movement	38.9	46.5	52.1

Shoulder Forces

The maximum push condition generated the highest moment of force at the shoulder joint (74.3 Nm). The shoulder was flexed to 30° from horizontal with the elbow in full extension. The forces transmitted to the shoulder directly related to the force required to move the breaker (Table 7-7). There are presently no standards available in the literature to indicate the level at which the shoulder is placed at risk of injury.

ERGONOMIC RECOMMENDATIONS

The ergonomists suggested an alternative means of handling the circuit breakers once the problem was quantified. This alternative required modifying the old breakers and cells by employing a racking lever to move the breakers in and out of the cell. The proposed racking lever system consisted of a long lever arm and foot attached to the underside of the breaker (Figure 7-3). Applying a downward force to the lever moves the foot along a tracking system on a removable ramp.

Eliminating awkward work postures required to push and pull the breakers is the mechanical advantage of the racking lever. As a result, the nature of the job is significantly changed so that little pushing or pulling is required. Instead, the force required to move the breaker becomes a much smaller vertical force (30 lb to 40 lb) exerted by the arm muscles acting on the handle of the lever. This change of force magnitude and direction substantially improved the worker's posture, thereby reducing the stress on the lower back.

The handles from the front of the breaker were removed to discourage attempts at manually pulling the breaker. In addition, the workers were taught the proper techniques to protect their backs while handling the breakers and other heavy loads. The task analysis videotapes were useful for illustrating breaker handling techniques.

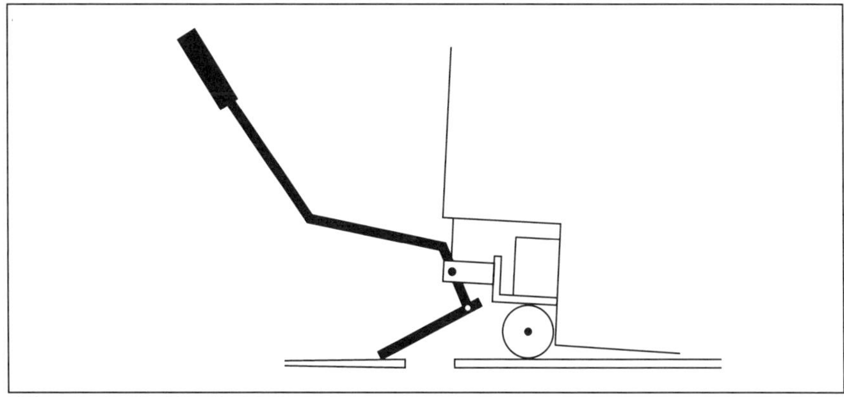

Figure 7-3. Racking Lever System.

Table 7-8. Estimated Injury, Production, Equipment, and Maintenance Costs.

Injury Costs			
Medical Costs	# of Visits	$ Per Visit	Total
old	12	64	$ 768
new	---	---	------
Compensation Costs	# Days	$ Per Day	Total
old	61	62	$3,782
new	---	---	---------
Lost Time Costs	# Hours	$ Per Hour	Total
old	322	20	$6,440
new	-----	---	---------
Estimated Total Injury Costs			
old			$10,990
new			----------

Production Costs:	Task Cycle Time			
	Unit	Yearly	Hourly Wage	Total Production Cost
old	130 sec	65 hrs	$20	$1,300
new	65 sec	65 hrs	$20	$ 433

Equipment Costs	Modification (Design/implementaton)
old	----------
new	$11,164

Maintenance Costs	Damage Costs/Year
old	$ 1,000
new	---------

COST-BENEFIT ANALYSIS

Cost justification of the changes was used to help substantiate the recommendations. In this case four areas were examined: injury costs, production costs, equipment costs, and maintenance costs (related to damaged equipment).

Four accident or injury reports related to moving circuit breakers were filed from May 1986 to September 1988. The short-term costs to the company were broken down into medical costs, compensation costs, and lost time costs (Table 7-8).

COST SUMMARY

The cost of modifying the equipment is approximately $11,200; whereas the estimated costs resulting from handling the breakers is $13,000. This estimate does not take into account the possible long-term effects of the injuries, the potential costs associated with reinjury, and the staffing problems presented by restricting injured personnel from this task.

FOLLOW-UP

Following the development of a prototype device, representatives from the human factors, electrical, and engineering departments met to give final approval of the design. All personnel that tried the new device were satisfied with its performance and design. There have been no medical costs associated with use of the modified system. The long-term impact of modifying the breaker task will be assessed through continued analysis of medical and accident records.

Trials with the prototype device showed a significant reduction in time to do the task. Implementation costs are controlled through in-house manufacture of the racking lever system.

Comparing the projected total savings with the cost of the recommendations clearly demonstrates the benefits of applying ergonomics to a difficult manual task. The most significant improvement both in terms of cost savings and worker well-being is the reduced potential for back injury as a result of the breaker modification.

References

1. Snook SH. The design of manual handling tasks. *Ergonomics.* 1978;21(2):963-986.

2. Norman RW. *WATBAK, A Computer Software Package for the Estimation of Low Back Compressive and Shear Forces.* Waterloo, Ontario, CAN: University of Waterloo; 1984.

3. National Institute for Occupational Safety and Health. *A Work Practices Guide for Manual Lifting.* Cincinnati, OH: U.S. Deptartment of Health and Human Services (NIOSH); 1981; Technical Report No. 81-122.

Jean M. Dalton is the coordinator of Human Factors and Rehabilitation at Dofasco, Inc. The group includes three Ergonomists, an Occupational Therapist and a Physiotherapist. Prior to working at Dofasco, Ms. Dalton worked teaching ergonomics and consulting. She obtained her BSc in Human Kinetics from the University of Guelph.

Nancy Smitten is an ergonomist with Dofasco, Inc. Her work involves assessment of jobs in an industrial setting with specialization in biomechanics problems. Her previous work experiences include ergonomics consultations for the mining industry and assessment/ treatment of injured workers. Ms. Smitten received her BSc in Co-op Kinesiology from the University of Waterloo.

Section III

Information Ergonomics

8

Methodology for Human-Computer Interface Design

C. Marlin "Lin" Brown, PhD

Too often, computer-based systems designers are experts in computer hardware and software, but know little about the capabilities, limitations, skills, tasks, and needs of the users. The results are systems that fail to adequately support the users' tasks, and require more effort and training to learn than users (or their employers) are willing to invest. Also, these systems may place new demands on users, making their tasks more difficult and time consuming rather than easier and more efficient.

INVOLVE HUMAN-COMPUTER INTERFACE EXPERTS

To design a useful and usable system it is important to involve people with skill and experience in user interface design. Almost any computer-based system will benefit from the design support of expert human-computer interface designers. User interface experts are typically human factors engineers, industrial engineers, cognitive psychologists, or software engineers who specialize in this field. They have a special interest in usability and in serving as advocates for the user in the design process.

INVOLVE AND KNOW THE USERS

Designing an effective user interface requires knowing who the users are, what they do, and how the proposed computer system can help users do their jobs more easily and quickly. Designers use several tools to gain an understanding of the users and their jobs, such as interviews, surveys, questionnaires, and market analyses. Because knowledge of the users is critical to successful system design, the users must be directly involved in the design process from its earliest stages, and work with designers to ensure the system meets their needs and is functional and usable.

Interviews, Surveys, and Questionnaires

Potential users of a proposed system provide one of the most valuable sources of information about the kind of tasks the system needs to support. In the early stages of a design project, designers should tap this source via interviews, surveys, and questionnaires to help understand how users do their jobs, how the proposed system could best help them, how existing systems work, and what are the strengths and weaknesses of the existing systems. As system prototypes are developed and refined, designers should demonstrate them to potential users, have the users operate them, and incorporate user feedback into subsequent iterations of the prototype.[1]

Users on the Design Team

People who are representative of typical users of the proposed system should be on the design team. They can provide valuable details of user tasks that can influence system design. Because they have been involved in system design, the user representatives can also serve as advocates to help ensure the system is accepted.

This does not imply that user representatives should have the power to dictate design details. Existing system users are often too familiar with old, well-learned systems and procedures to recognize the value of some improvements. Their opinions, experience, and job knowledge, however, should be represented and respected on the design team.

Market Analyses

When other systems or products provide similar functions, address similar markets, or support similar tasks as the system under development, they can be fertile sources of information. Which ones are most successful? What do users like and dislike about them? Which ones are easiest to learn, and why are they easily learned? Which ones are most efficient for experienced users and why? What are the generally recognized problems with these systems? How can these problems be avoided?

ANALYZE USER TASKS

The design must be based on an understanding of the tasks the users will perform with the system and the physical and sociological environment in which it will be used. A task analysis is a time-oriented description of the user-system interactions, which will show the sequential and simultaneous manual and intellectual activities of users, rather than just sequential operation of the equipment. Two kinds of task analyses can provide valuable, systematically derived insights to guide system design: analysis of users' existing tasks without the benefit of the new or proposed system, and analysis of the tasks required for operating the proposed system.

When analyzing user tasks, designers should systematically identify the tasks and subtasks to be performed. Also, the designers should identify and describe the *task flows*, defined as the order in which tasks are performed, and the dependencies of tasks on preceding tasks. Bottlenecks and error-prone subtasks should be identified and addressed in the subsequent system design.

Task analyses are also valuable for allocating functions, both to automatic (computer) versus human processing and among various operators or user tasks. Often a new system will lead to changes in the number of people required, their skill requirements, and the nature of their jobs. Results from task analyses can be used to specify the number of job categories (positions), the number of people required in each job (staffing), and the skills required by the people at each position. When analyzing existing tasks to project new tasks, designers should pay particular attention to workload at various job positions and the level and types of coordination required among users.

DEVELOP AND UTILIZE DESIGN GUIDELINES

Human-computer interface design guidelines describe the conventions and practices for designing user interfaces that promote consistency, ease of learning, and ease of use of the resulting computer-based product. Guidelines address such topics as the design of dialogues between users and computers, display formats, control and display devices, error procedures, the use of color and graphics, and effective techniques for wording and coding.[2] The following discussion emphasizes the value of guidelines in the design process, and suggests strategies for developing guidelines tailored to the environment as well as to the constraints of the design project.

Importance of Design Guidelines

Project-specific user interface design guidelines should be developed, documented, revised, and maintained. The general guidelines in the literature[2,3] are good starting points, but they must be tailored to the project.

Human-computer interface guidelines document the concepts and agreements among designers about the project's user-interface conventions. The requirement to establish and document these conventions as guidelines leads designers through several beneficial steps. In the process of establishing a mutually accepted set of conventions, many important design issues and trade-offs will be uncovered and addressed early. The impacts of proposed user interface approaches on diverse design aspects, such as user needs, programming, hardware design, maintenance, testing, manuals, access control, and user support, are often revealed, leading to more balanced, optimized trade-offs. Documenting conventions promotes consistency in the user interface in spite of the fact that many different designers, each with different personal preferences and styles, are developing different modules or components.

Role of General and Specific Guidelines

If guidelines are to be useful, they must present specific, relevant design rules. Platitudes and vague advice are of little value to a designer. A good test of the guidelines' quality is the extent to which they serve as audit criteria. For example, can designers readily determine if a given interface design meets a guideline? If not, the guideline probably needs to be stated more clearly.

In a large computer-system program several autonomous design groups often work on major subsystems. In these situations, general guidelines help ensure consistency and ease of use at the system level, especially if the same users will access more than one of the subsystems. General guidelines, however, usually cannot provide the level of specificity necessary to serve as detailed design conventions without unnecessarily constraining subsystem options. Local design convention documents, developed for each subsystem can provide detailed, tailored guidance for the special needs, constraints, and users of each subsystem.

Guidelines also need to be realistic and reflect the hardware and software constraints of the system for which they are written. A few unrealistic guidelines can damage the credibility of the entire set. In other words, if general guidelines are applied in a specific setting with known constraints, they should be tailored to be realistic within those constraints.

Benefits to users. The following are two examples of how guidelines can benefit users. Keister and Gallaway[4] redesigned the user interface of an existing commercial software package to make it conform to several commonly accepted human-computer interface guidelines. They then compared the actual data entry performance of the original and revised programs. Those using the modified programs completed tasks 25% faster and made 25% fewer errors than users of the original programs.

Burns, Warren, and Rudisill[5] revised the formats of displays used by space shuttle astronauts and flight controllers to meet accepted human-computer interface guidelines. The revised displays improved the performance of novice users, as well as the astronauts and controllers who were well-trained experts in the original displays.

Benefits to design projects. Although guidelines primarily benefit users, several side effects also benefit design projects. Guidelines promote the use of standard procedures, which tend to provide a predictable, familiar user interface. Also, standard procedures provide systems analysts and programmers with a familiar, common frame of reference for developing the user interface. By providing a systematic, documented framework for user interface design, guidelines can simplify many design problems. They can even head off some problems before they occur, so that no valuable design time is wasted.

The consistency of user interface design, which guidelines foster, often permits programmers to use the same programming code or modules in many applications. This dramatically reduces the amount of unique user interface code to be developed for each module.

A documented set of standard user interface procedures and conventions can be of immense value to programmers, systems analysts, and others who are new hires or transfers to an ongoing project. General guidelines, and especially local user interface conventions, make effective training materials.

USER INTERFACE

The user interface design involves some key steps, each of which are important to the success of the design. The following discussion presents some of these steps.

Functions (features) Recommendations

How useful the system is and how easy it is to use should be key considerations in defining the system's functions and the user interface. Human-interface specialists should participate in identifying, describing, and documenting these functions. This step is essential to establish *what* the system is required to do before proceeding to *how* it will do it. The functions should be based on an analysis of competing products, users, tasks, human interface research, technology surveys, trade-offs, and feasibility.

Industry Standards

In some design situations, industry standards may be relevant to the design of the system, and may be very helpful in determing the fundamental aspects of the user interface look and feel. For example, several *de facto* standards are evolving that describe standard rules for mouse usage, menus, screen windows, on-screen buttons, and help systems. When considering these standards, designers need to evaluate the benefits of using a particular standard and decide which one(s) is/are the best for that particular cost. This evaluation will require an analysis of the specifications, advantages, and disadvantages of each relevant alternative.

Dialogue and Display Appearance Design

Dialogue design defines the ways in which users can interact with the computer system. Common dialogues are menus, commands, form filling, and function keys. The choice of dialogues and the ways in which they are implemented are perhaps the most critical factor that determines ease of use.

The appearance of a display screen represents a series of design decisions. What information does the user need to perform the task? At what level of detail? What formatting, grouping and highlighting techniques make that information accessible? What corollary information should be readily accessible?

Effective and usable displays should be based on a detailed analysis to support these decisions. The evolution of high-resolution graphic displays has enabled the use of far more sophisticated representation and pictorial elements. As a result, attention to detail in the display format design is even more critical.

Design and Specification

The display and control device hardware by which the user communicates with the computer system, must be carefully selected and designed. The selection of the appropriate input devices, such as the alphanumeric keyboard, mouse, touch sensitive panel, and pushbuttons; and display devices, such as video monitors, printers, and audio displays, should be based on the tasks, the users, and the environment. Ergonomic factors, such as those that might affect eyestrain or muscular strain, anthropometric factors, such as reach, control separation, work-surface height, and control size: and aesthetic factors must be included as design criteria.

As a user interface design develops, it will gain visibility if it is formally documented in a design specification. In addition, the process of writing the specification will typically reveal any areas that have not been sufficiently defined. The specification will also serve as a means for communicating design concepts.

ITERATIVE DESIGN

Prototypes, user testing, and redesign are the three key components of the iterative design process. Developing prototypes permits original design concepts and redesigns to be turned into something concrete. The prototypes make the design more understandable to the users, forces the designers to work out the details, and serves as concrete proof of the designers' abstract concepts. Testing of prototypes promotes an active user role in system development, user commitment to the system, and user acceptance of the system.

Prototypes

A *prototype* is an early version of a system that exhibits some key features planned for the later operation system. Design guidelines and preliminary design specifications cannot foresee all of the user-specific, task-specific, or organization-specific implications. Prototypes provide a way to identify and refine poorly defined requirements so that design problems can be detected and corrected earlier in the development process and at less cost.

Working prototypes permit testing through operation use. They enhance understanding of the system by permitting users to experience the implications of design concepts firsthand. Also, they facilitate user-designed communication.

Additionally, prototypes can provide real data for use in estimating the resource requirements of the operation system, such as development cost and schedule, response time, and computer capacity requirements, as well as an effective and demonstration tool. Using prototypes often requires less time and money than traditional approaches. In some cases, the final prototype iteration may become the system.[6]

Paper and pencil. Long before any coding of system software has begun, early user-interface concepts can be tested by providing users with paper simulations of the interface design. Examples include simulated user manuals, system-usage scenarios, paper simulations of screen formats, and paper-and-pencil tasks. These simulations represent certain critical aspects of the user interface, and can be used to test and refine concepts and plans.[1]

Wizard of Oz technique. By simulating planned systems, more sophisticated tests can be conducted before an operation prototype is available. A hidden human user, for example, can simulate the responses that a planned but nonexistent system would make. Gould, Conti, and Hovanyecz[7] used a simulation to study user performance on an automated dictation system, or listening typewriter, in spite of the fact that speech recognition technology does not yet allow unrestricted dictation to a computer. People spoke into a microphone, and their dictated words were displayed back to them on a CRT. Unseen by the people was a typist in another room who listened to the dictation and typed the words into the computer.

This testing method has been named the *Wizard of Oz* technique.[8,9] Gould and his colleagues were able to make valuable conclusions about the effects of vocabulary size and connected word versus isolated-word speech through this simulation, even though the technology to support automatic dictation does not yet exist.

Mockups and models. Mockups and models are physical facsimiles of planned devices or designs that are used to visualize, demonstrate, and evaluate the external appearance of a proposed or actual system. A full-scale facsimile is typically called a mockup. A model is typically smaller than a full scale facsimile. Mockups and *models* are useful, because important details that may be less apparent in drawings often become obvious. They are particularly useful in validating dimensional assumptions about ease of use, safety, and comfort; for example, previously undetected problems in reach, viewing angle, and access to controls.

Video tapes. Video taping is a useful way to simulate proposed user interface designs. Static displays can simulate dynamic displays, using splicing and other video production techniques. Photographs, drawings, and other paper images can be video taped to produce a realistic-looking simulation of displays and devices. Video is also a portable medium. A prototype can be demonstrated to potential users or customers in other locations without having to transport it.

User Testing

Designers must test design features and concepts on people for whom the system is targeted. Key concepts can be tested early in the design process by using paper simulations of formats, dialogues, and messages. Subsequently, prototypes of gradually increasing scope and fidelity can be tested on users. The lessons learned from each test iteration can be incorporated into the next design iteration.

Preliminary testing can use design team members, friends, or fellow employees but that does not substitute for testing users from the system's target population. Tognazzini[10] emphasized the importance of testing actual users:

"Products must be repeatedly tested on 'real people.' ('Real people' means the target audience: as soon as you find yourself sitting in a meeting with other computerists, all announcing what users will or will not feel/think/do, you are in trouble—build the prototype and find out.)"

Testing should measure users' behavior objectively, not just their opinions. Designers should observe and measure the performance of users operating the simulated or prototyped system. Reviews and demonstrations alone do not require users to deal with the details of system operation firsthand, so the opportunity to collect valuable diagnostic data may be lost. Empirical data should be systematically collected, including learning time, the number and types of errors made, the time taken to complete tasks, retention, and the users' preferences and satisfaction with aspects of the system.[1]

Redesign

The lessons learned from testing, simulations, and prototypes are important, and should be incorporated into a revised system design. Revision and retesting of the design by measuring user performance should be repeated as many times as necessary. This process will make the development cycle longer and more costly, but early testing and iterative design will usually speed overall development by detecting and correcting problems early. Prototypes and design iteration can also help get a project going by stimulating thought, providing an embodiment of design plans, and providing something tangible to show others.[1, 10]

FOLLOW-UP EVALUATIONS

Once a system has been designed and fielded, follow-up evaluations permit designers to check the validity of their assumptions about its usefulness. Since most systems are occasionally upgraded with new software or hardware releases, a product should never be considered final.

User assistance consultants can be a valuable source of information for evaluating and improving a fielded system. Automatic collection of evaluation data can be built into the system by techniques, such as an online suggestion box or automatic tallying of the frequency of specific errors, help requests, commands, and menu requests. Error patterns may reveal dialogue elements that cause problems for users, and should be redesigned in future releases.

Usage patterns may show that some parts of the system's functions or elements of its dialogues are used extensively, while others are rarely if ever used. The seldom-used functions or dialogue elements should perhaps be eliminated or receive less of the revision effort in future releases.

Even if subsequent releases are not anticipated, the investment in a follow-up evaluation may be well-spent. The design principles and assumptions on which the fielded system was based should be validated, since designers may carry many of these concepts with them into their next project.

References

1. Gould JD and Lewis C. Designing for usability: key principles and what people think. *Communications of the ACM*. 1985;28(3):300-311.

2. Brown CM. *Human - Computer Interface Design Guidelines*. Norwood, NJ: Ablex Publishing; 1988.

3. Smith SL and Mosier JN. *Design Guidelines for User-system Interface Software*. Bedford, MA: MITRE Corporation; 1984; Technical Report ESD-TR-190.

4. Keister RS and Gallaway GR. Making software user friendly: an assessment of data entry performance. *Proceedings of the Human Factors Society 27th Annual Meeting*. Human Factors Society; 1031-1034.

5. Burns MJ, Warren DL and Rudisill M. Formatting space-related displays to optimize expert and nonexpert performance. In *Proceedings of CHI '86 Human Factors in Computing Systems*. New York: Association for Computing Machinery; 1986;274-280.

6. Alavi M. An assessment of the prototyping approach to information systems development. *Communications of the ACM*. 1984;27(6):556-563.

7. Gould JD, Conti J and Hovanyecz T. Composing letters on a simulated listening typewriter. *Communications of the ACM*. 1983;26(4):295-308.

8. Kelley JF. An empirical methodology for writing user friendly natural language computer applications. In *Proceedings of CHI '83 Human Factors in Computing Systems*. New York, NY: Association for Computing Machinery; 1983;193-195.

9. Green P and Wei-Haas L. The rapid development of user interfaces: experience with the Wizard of Oz method. *Proceedings of the Human Factors Society 29th Annual Meeting*. Human Factors Society; 1985;470-474.

10. Tognazzini B. *The Apple II Human Interface Guidelines*. Cupertino, CA: Apple Computer; 1985.

C. Marlin Brown is Manager, User Interfaces, at Independence Technologies, Inc., where he is responsible for the design of user-system interaction dialogues for computer work stations, point-of-sale transaction terminals, and telephone voice response units. He has developed system concepts and user interface designs for a variety of systems including a computer-supported pulmonary testing lab, a medical data base for cardiovascular patients, corporate data bases for finance, engineering and material, and space systems, including a mission control center for the space shuttle.

Dr. Brown received his PhD in Engineering Psychology from the Georgia Institute of Technology. He is currently Director for the Bay Area Chapter of the Human Factors Society, and has served as President and Vice President in previous years. He is also a member of the IEEE Computer Society and ACM Special Interest Groups on Computer-Human Interaction (SIGCHI) and Artificial Intelligence.

9

A Job Performance Aid for AS/RS Error Diagnosis and Scheduling Preventive Maintenance

B. Mustafa Pulat, PhD
Timothy M. Bryan

Maintaining systems and machinery at maximum levels of uptime with minimum cost is one of the most important missions of any enterprise, especially one operating in a capital-intensive environment. National surveys, however, show that few companies are achieving this objective.[1]

The *Maintenance Engineering Handbook* [2] groups maintenance into two general classifications: primary functions that require day-to-day work, such as maintenance of existing equipment, buildings and grounds, inspections and new installations; and secondary functions, such as waste handling, salvage operations, and storekeeping.

JOB PERFORMANCE AIDS

As systems and machinery become more complex, human skills and abilities become less reliable in coping with the operation. In these instances, job performance aids (JPAs) extend human capabilities, although at some cost. Due to their complexity, maintainence tasks have been the target for most JPA development.

A *job performance aid* is a device or document that provides needed information to carry out a task effectively.[3] Often, JPAs help workers store and maintain information, make the information available on the job, and reduce the amount of decision making necessary to perform a task by providing step-by-step guidance to a probable solution or cause. Job performance aids are most helpful for tasks that involve calculation, stringent memory requirements, accuracy, difficult decisions, and multiple judgments. They are also appropriate for repetitive and boring tasks.[4]

The following case study illustrates how a job performance aid in the form of a computer system helped a maintenance crew be more effective. The crew's job is error recovery in a multimillion dollar automated storage and retrieval system (AS/RS) at the AT&T Oklahoma City Works. Because hundreds of items are moved to the manufacturing shops from the AS/RS each day, it is extremely important that the system's throughput capacity be maintained so the downstream manufacturing operations would not be affected.

THE AUTOMATED STORAGE AND RETRIEVAL SYSTEM

In November 1985, the Works began a miniload AS/RS that operates in the buffer zones between central receiving and the manufacturing shops (Figure 9-1). The system automates the in-plant function of receiving, stocking, and dispatching of piece-parts material management. It also handles a variety of audit and administrative functions. Sophisticated laser scanners, sortation mechanisms, conveyors, and software processes control the movement of the material in self-contained form or containers (totes, bins). Two AT&T 3B20A processors, one as production the other as warm standby, are the primary system controls.

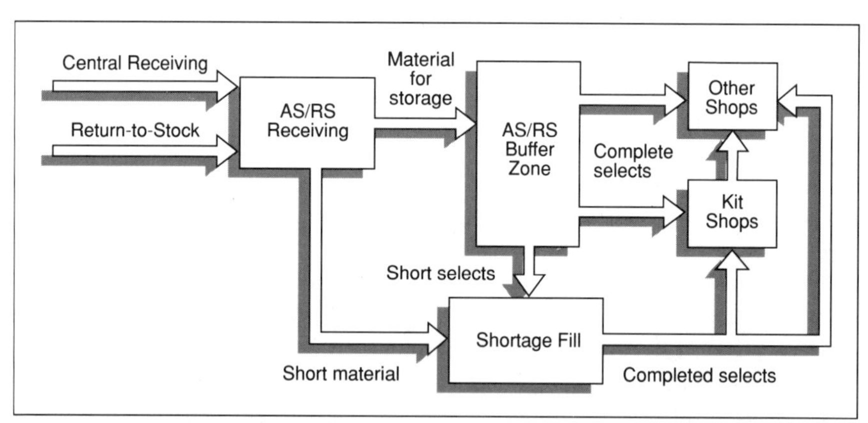

Figure 9-1. AS/RS Material Flow.

The major hardware components are the storage and retrieval (S/R) robots, the rack structure, and the pick and delivery (P/D) stations. Aisle captive S/R cranes operate in nine aisles. The rack structure is 30 ft tall and 80 ft long and displays over 24,000 bin locations. The cranes are single-masted with one lateral shuttle to store or retrieve loads on each side of the aisle. The shuttle can monitor and reject loads that exceed predetermined dimensions. The mast is supported on a track and guided at the top by rails for accurate vertical alignment. Electrical power is supplied to the crane by a power rail mounted at the bottom of the aisle. A deceleration sensor brings the crane to a safe and smooth stop before it contacts the mechanical end stops.

Crane Fault Conditions

In terms of maintenance attention, the crane is the most critical element in the system. Its moving parts, mechanical interfaces with other hardware, such as the rack structure and power and guide rails, its communication interfaces, with the host computers and manned stations, make the crane highly vulnerable to downtime. Once a crane is down, activities such as stocking, retrieval, and auditing cease in that aisle.

Each crane's work is downloaded from the material requirements planning (MRP) system throughout the day. System administrators can also schedule retrieval orders. The crane is under computer control during stocking or retrieval as long as it encounters no electromechanical problems. If there is a problem, such as pins jammed or a bad limit switch, the crane control is switched automatically to manual mode and the crane stops.

Prior to the development of the JPA, when an error condition was detected, the stocking operator followed a manual quick-fix procedure, based on job knowledge (Figure 9-2). Because most error conditions are complex and the associated memory requirements of the operator are heavy, the quick fixes usually failed. By the time the AS/RS maintenance group was informed, significant downtime had occurred and, in some cases, the original fault condition became a complex problem.

The original fault diagnosis and correction process did not allow recording of any historical data of errors made by the crane. Such data are useful in pinpointing root causes of faults and may aid in quicker error recovery.

Consequently, the process was obviously incapable of consistently maintaining high levels of crane uptime. A JPA was necessary, which would allocate the function of fault interpretation for over 70 possible fault conditions to a mechanized process. The JPA also would present simple suggestions to the maintenance crew for the best error recovery scheme for each fault condition. Furthermore, the process could automate the communication between the crane and the maintenance crew in addition to archiving error condition data.

DEVELOPING A JPA

Representatives from the Works' material handling engineers, ergonomics specialists, information systems developers, the AS/RS software developers from AT&T Bell Laboratories in Columbus, Ohio, and relevant personnel from the crane vendor all participated in the JPA development process. Protocols for passing information between the AS/RS software and the crane logic unit software were established and shared memory locations were designed for instant on-line reporting.

Figure 9-2. The Diagnostic Panel and a VDT Are Used for the Quick-fix Procedure.

Special software was developed to translate data into informative reports. The crane field engineer was interviewed to develop accurate rules for error recovery. These rules were then mapped into a rule-based system. Hardware was provided to the maintenance crew and other relevant personnel using the system. A main menu guides the maintenance crew to the appropriate JPA function. The five major functions are as follows:

Interactive error recovery. This pre-expert system forms the most important module (Figure 9-3). It takes the user (maintenance crew member) through a fault condition to suggest the most effective means of recovery, which may require adjustments on any combination of over 400 crane parameters in the crane logic unit software. The driver here is the rule-based knowledge system. For the more experienced users, the JPA allows for graphical, fast step-through.

Preventive maintenance schedules. The software maintains and issues all required preventive maintenance schedules on request. The maintenance crew check this function daily to work on vital components before problems arise.

Figure 9-3. Interactive Error Recovery Guides the User to the Best Corrective Action.

Request maintenance. A special request is documented. A section chief may request maintenance work on any part of the AS/RS using this option. The request is automatically printed on the maintenance printer, logged into the computer records, and a copy is forwarded to the material handling engineer. The request is assigned a number that must be cancelled by the engineer after work completion.

Print special reports. Reports can allow an analysis of the error condition using performance records of any crane (Figure 9-4). In many cases, this capability is invaluable in pinpointing the root causes of crane problems; for example, downtime statistics such as error codes, frequencies, and durations by the minute, by day, and by crane. A special instruction list is also available to perform certain duties such as putting the crane in random mode for testing. Higher-level charts can also be generated to review overall crane performance.

Monitor bar code scans. The maintenance crew can monitor the performance, such as misreads, of all bar codes scanners for potential preventive maintenance.

CONCLUSIONS

In many situations, human performance must be supported by JPAs, tools, equipment, and other means for adequate system performance. Workers cannot be expected to function effectively under sensory and mental underload and overload.

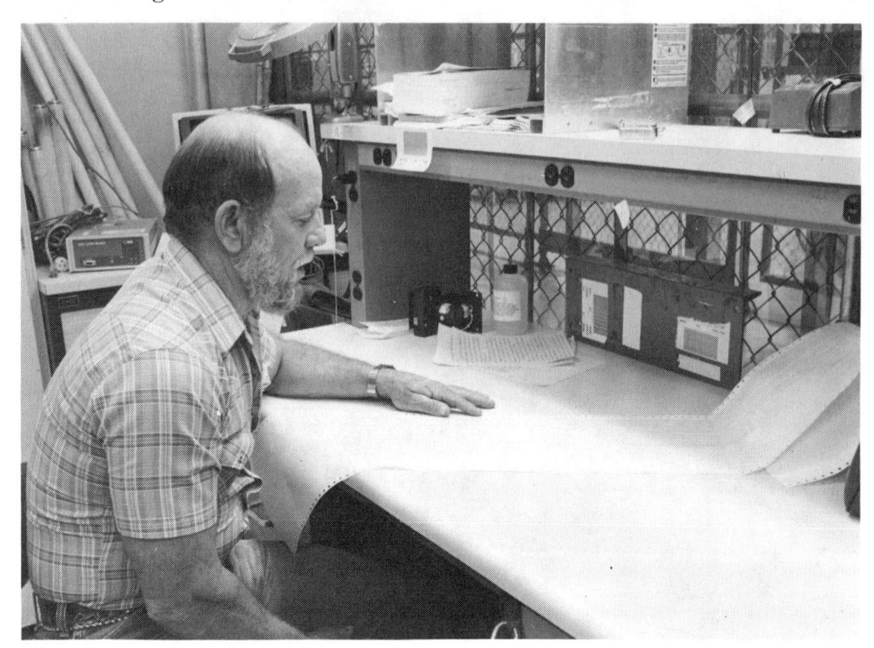

Figure 9-4. Many Reports are Available From the System for Best Maintenance Decisions.

Because of the JPA, AT&T Oklahoma City Works is noticeably reducing the crane downtime and increasing the direct labor hour savings at the AS/RS. The crane downtime percentage has been reduced fourfold with well-controlled variability. Furthermore, a 50% reduction in labor hours has been observed, with 99% reduction in the engineer's attention time.

Also, the JPA was not imposed on the AS/RS operators or maintenance crew. They do not feel they are trained robots who can only perform the simplest tasks. A valid business need voiced by the operating personnel triggered this development. As a result, the JPA was accepted by everyone.

References

1. Higgins LR and Morrow LC, eds. *Maintenance Engineering Handbook*, 3rd. ed. New York, NY: McGraw-Hill; 1977.

2. Nolden C. Plant maintenance costs. *Plant Engineering.* 1987;41(14):38-42.

3. Rifkin KI and Everhart MC. *Position Performance Aid Development.* Valencia, PA: Applied Science Associates; 1971.

4. Swezey RW. Design of job aids and procedure writing. In: Salvendy G. ed. *Human Factors Handbook.* New York, NY: John Wiley and Sons; 1987;1039-1057.

Timothy M. Bryan is an engineering manager at the AT&T Network Systems, Warrenville, Ill. He oversees the project management and planning work associated with the PRISM (Productivity Improvement Systems for Manufacturing) software control systems that are deployed at the Network Systems manufacturing facilities.

Mr. Bryan obtained his BS and MS degrees in industrial engineering from Oklahoma State University and Purdue University, respectively. He is a member of the Institute of Industrial Engineers and has published in the areas of productivity improvement, systems engineering, and ergonomics.

10

Software Improvements Enhance Throughput in the Raw Material Buffer Zones

B. Mustafa Pulat, PhD
Jack L. Glover

Effective material movement throughout the business cycle is an important manufacturing concern, and depends heavily on delivering the correct products or materials to the right place at the right time at minimum cost.[1] With material and handling costs ranging between 50% to 80% of product cost, the electronics manufacturing industry is particularly sensitive to this concern.

Efficient material movement requires coordination, which becomes even more important as the material handling function is automated. Fast, timely completion of these activities frequently requires complementing the physical flow of materials with a parallel flow of information. This information flow provides the basis for many material flow control decisions such as routing and volume control.

THE HUMAN ELEMENT

Although many material handling functions are automated, the human element is crucial, even in highly automated systems. People play important roles at the infeed and discharge points. Futhermore, people are still the best decision makers. Computers usually monitor complex handling systems such as an automated storage and retrieval system (AS/RS), but people are still in control of the total process, including monitoring the computers, initiating actions to recover from errors, and handling special conditions that cannot be automated. Consequently, system designers should integrate people into the design process as early as possible. System designers must be aware, however, of workers' resistence to change. The ergonomics literature provides the most comprehensive review of this aspect of system design.[2,3]

EXTERNAL BUFFER ZONES

This case study of the external buffer zones at the AT&T Oklahoma City Works demonstrates the need to have humans engineer automation projects from the start in order to integrate workers into the process as early as possible. The case study also illustrates specific ergonomic applications in an information automation project.

The *external zone* terminology was coined at the Works in 1986 to describe material buffer locations outside the physical limits of the mini-load AS/RS, which started operating in November 1985. These zones include external warehouses, high rack and shelving zones, and the unit-load AS/RS, which became fully operational in December 1988 (Figure 10-1). The unit-load AS/RS was designed to span most of the coverage of high racks, shelving zones, and the warehouses.

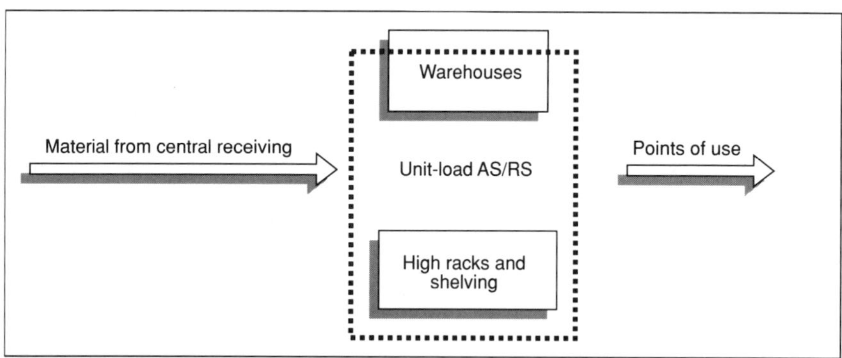

Figure 10-1. External Zones Material Flow.

The external zones buffer material, such as frames, fans, fan housings, and brackets, are purchased and handled in bulk primarily on pallets (Figure 10-2). Material routing decisions are made at central receiving. An incoming part to be used in the manufacturing process can be routed to several destinations, including the assembly shops, mini-load AS/RS, and any location within the external zones. Items to be buffered prior to manufacturing, are stored in a bin or rack. They are picked and dispatched to the point-of-use after receiving a pull signal or material requirements planning-manufacturing planning software (MRP) request.

In addition to the hardware, such as fork lifts, conveyors, and rack structures, efficient material handling requires software support. Basic storeroom functions, such as receiving, stocking, inventorying (counting), withdrawing, and accumulating, require the initial support. These functions then interface with other

Figure 10-2. Bulky Parts Are Moved Throughout the External Zones.

systems such as factory receiving and the MRP. Finally, all systems are integrated for a closed-loop information flow (Figure 10-3).

The primary reason for tracking material in the buffer zones is to give the production planning system an accurate view of on-hand balances. Production activities for existing orders and purchasing decisions for planned orders rely on such information.

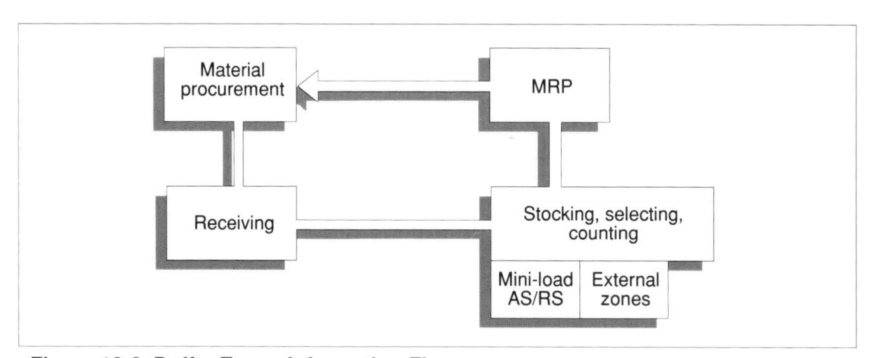

Figure 10-3. Buffer Zones Information Flow.

ERGONOMIC INTERVENTIONS

In 1985, the Works contracted the task of developing the necessary software support of external zones, including MRP interfaces, to an outside group. The software was scheduled for installation in February 1986. Two weeks of trials led management to conclude that this software needed a major revision. It was removed from use for five months, during which storeroom operating and engineering personnel worked closely with the developers to modify it.

Initial Release

The major problem with the first release was that the software did not integrate the physical movement of material with the information flow it supported. Many normal shop floor operations were not being recognized, which meant that material moved without an accurate record of the move. Naturally, this would have caused many accounting problems. Data records would be inaccurate, leading to erroneous MRP-initiated transactions. Other problems included:

- *Many partially supported functions*: Select cancelling, for example, was possible only by MRP transaction codes. Workers were also required to cancel selects by AS/RS (the execution software) ID.
- *Notable use problems*: A window withdrawal required signing-up to two different systems with 11 screens to be processed. On the average, five different screens had to be processed to track a single material move. The external zones processed thousands of transactions per day, and workers were overwhelmed by the system's requirements (Figure 10-4).
- *Unneccessary control*: Devout tracking requirements introduced many in-transit queues, which were accompanied by unnecessary paperwork.
- *Common system response times of 30 seconds or more*: This led to much worker frustration, because daily work could not be completed in time. Unnecessary overtime work would lead to significant costs.

Improved Release

Both the material handling engineering and material storage and distribution (operating shop) departments worked with the developer to improve the software. To close the operation/support compatibility gap as well as to solve user problems, the system functions were ergonomically screened.[4-6] Frequent visits between the Works and the developers facility in addition to electronic communications and intermediate trials resulted in an improved release of the software. This release was installed in June 1986. There were four areas of specific improvements.

New function support. In this category, improvements focused on providing support to those functions that were not initially supported. A detailed task analysis of the external zone operator functions, and a comparison of the results with those being supported by the first release, revealed the differences, and many of the new functions were realized by adding data fields to existing screens.

The advent of AS/RS ID is an example of the new function support. This five-digit serial number represents the MRP logical store, part identifier, description, assembly shop destination, and the required quantity. The AS/RS ID allows workers to make only one string entry to identify the attributes it represents. Other features include:

- Supporting MRP-generated counts
- Providing a new screen for periodic reviewing of AS/RS ID status
- Supporting return-to-stock receival type
- Developing a withdrawal option
- Allowing transfer of material from internal to external
- Allowing alien (not reported to the planning systems) parts to be stored in external zones

Figure 10-4. Operators Spend Many Hours at VDT Work Stations for Material Move Transaction Processing.

Improving support for existing functions. This group of improvements was aimed at enhancing coverage and effectiveness of functions already being supported, as well as eliminating those that were unnecessary. These improvements also allow in-transit tracking of warehouse material for location updates after stocking. The in-transit record is automatically canceled after a proper location update transaction. Some examples are:

- Eliminate most in-transit requirements
- Support external pick request printing on demand

- Eliminate comcode (part number) check for alien parts
- Enhance dispersal options
- Allow select cancelling by AS/RS ID
- Allow queues of selects to be canceled
- Permit overdelivery for bulk material
- Allow growth in external database
- Eliminate travel document screens
- Eliminate MRP reporting for alien parts
- Update the MRP with new count versus change in count

Improving user friendliness. By improving user friendliness, the goal was to allow speedier and error-free data entry with associated reductions in worker frustration. A secondary objective was to make the task of picking more efficient; for example:

- The external pick request form displayed sufficient locations for the total quantity to be picked and sorted by date of stocking. This made the picking task much more efficient and also allowed for first-in-first-out (FIFO) processing.
- Multiple transactions for the same part on the same screen became possible.
- External location codes represent unique physical zones; for example, all E codes represent warehouse locations.
- Extensive use of function keys allow for efficient transaction processing by either combining screens or allowing fast access between screens. Function keys were used to move from one screen to the next, with no need to go back to the main menu. Furthermore, a log-off transaction can be performed by one keystroke (the log-off function key).
- Other uses of function keys include a help function key, which displays help messages at the bottom of the screen concerning the field pointed out by the cursor. On the screen, users could use function keys to move one field at a time, redraw screens, clear a field, and display material locations one page at a time up to six locations per page.
- All function key images appear at the bottom of the screen at consistent locations. Such cues minimize error potential.
- Many other internal checks, such as part number, location, and AS/RS ID, assure data integrity and negate human-initiated errors.
- One screen allows for several types of receivals to be made on the same screen, such as new item, return-to-stock from shortage, return-to-stock from shops, and restock.
- All locations for a part are displayed across the external zones.
- For quantity updates, both the original and the new quantity values are displayed for sanity checks.
- Whenever possible, workers use the space-bar key to cycle through

allowed options for an entry. The option list is sorted by the expected frequency of selection.
- At the point of receival, workers use a one-screen, automated access to two different systems to perform an electronic check of the MRP shortage requirement, which fills shortages.
- Just-In-Time (JIT) material handling via MRP is performed the same way. There is no need to access MRP screens for this function.
- Guidance messages appear at consistent screen locations, such as "no more parts at this location" at the top of the page.
- Certain screen prompts change automatically depending on the function performed, such as "Quantity ——" with "stocked" or "dispersed" as possible automated selections.

Procedural improvements. We also focused on procedural improvements within and between functions. In addition, certain help functions were considered for operational effectiveness. The following are examples:

- Supporting an external transaction report, such as an activity on parts by day by worker
- Mass cleaning of in-transit quantities
- Printing of external pick request reports on high-speed printers
- Permitting access to the external main menu from internal main menu and vice versa
- Displaying the material movement screen automatically after a receival with nonzero entries in certain fields; otherwise the receival screen is redrawn. In many cases, an automated move from one screen to the next also retains information for relevant fields on the next screen.
- Following an orderly, logical sequence for reporting material move at every stage. For example, on the external material movement screen, the item to be moved is identified first, then information concerning all locations accessed to pick the part is entered. Next, confirmation of the pick operation is requested, after which dispersal information is gathered.

CONCLUSIONS

The major reason why the initial release of the external software failed was ineffective developer-user communication. System requirements were not established clearly, with little understanding of human characteristics. In the redesign, a two-pronged approach centered on a value-added approach aimed at clear definition of the necessary functions to be supported, and the user-friendliness approach focused on engineering the human interface points for improved use characteristics consistent with human expectations.[7]

The improved system has been operating for more than two years with minimal complaints. A window withdrawal process spans two screens, equivalent to an average transaction screen span. System response time improved more than 70%, which was also helped by attached processors. Operators are satisfied with the new release.

This case is another example of the need to develop systems with the user and not for the user. Close cooperation with the user is necessary for effective system development; integration of humans in the development process is a must for any system to succeed.

We now have a formal user acceptance testing (UAT) process that we exercise before installing any system. Those that do not pass our stringent screening get reworked until both engineering and operating departments are satisfied.

References

1. Kulviec RA, ed. *Materials Handling Handbook.* New York, NY: Wiley-Interscience; 1985.

2. Alexander DC and Pulat BM, ed. *Industrial Economics: A Practitioner's Guide.* Norcross, GA: Industrial Engineering and Management Press; 1985.

3. Sheridan TB and Johannsen G. *Monitoring Behavior and Supervisory Control.* New York: Plenum Press; 1976.

4. Galitz WO. *Humanizing Office Automation.* Wellesley, MA: QED Information Sciences; 1984.

5. Grandjean E. *Ergonomics in Computerized Offices.* London: Taylor and Francis Ltd; 1987.

6. Ivergard T. *Information Ergonomics.* Sweden: Chartwell-Bratt, Ltd.; 1982.

7. Didner RS. A value-added approach to information systems design. *Human Factors Society Bulletin*; 31(5)1-2.

Jack L. Glover is currently a section chief in the material storage and distribution center at the AT&T Oklahoma City Works where he manages the various operational units including the two automated storage and retrieval systems.

Section IV

Design of Work Space and Work Methods

11

Industrial Work Station and Work Space Design: An Ergonomic Approach

Biman Das, PhD, PE

Industrial work station design ideally should satisfy the system's performance requirements as well as the needs of the human user. Generally, an industrial workstation is a small working area, in which workers perform assigned tasks. The terms *work station*, *work space*, and *work environment* are often used interchangeably and include chairs, tables, machines, tools, actual product, regular and protective clothing, lighting, and climate; everything except the worker.

The physical dimensions of the work space are very important because small changes can have considerable impact on worker productivity and occupational safety and health.[1] The design, for example, should enable the worker to see the working area clearly, posture must be adequate and comfortable, and controls must be within reach to minimize error. Inadequate posture from an improperly designed work station can cause static muscle effort, which eventually results in acute localized muscle fatigue, decreased performance and productivity, and increased possible worker-related hazards.[2, 3]

The designer of a work station must consider the physiologic, psychological, environmental, and dimensional factors, which will affect worker performance and well-being. No doubt the overall design will depend on the interaction among these factors. The physiologic factors, for example, will have a direct impact on the worker's physical comfort. An optimum industrial work station must provide adequate body support, proper distribution of body-limb weight, natural body-limb positions, and should require little demand to use maximum reach or force.

The main psychological objectives of worker acceptance and motivation can be achieved if the workstation is simple, convenient, attractive, reliable, and safe. Environmental factors, such as illumination, temperature, ventilation, noise, and vibration, must also be taken into account. The worker-related dimensional factors that influence industrial work stations include:

- Posture control and distribution of body weight
- Reach envelope of hands
- Eye position with regard to display area

Additionally, work station dimensions should be consistent with anthropometric characteristics of the workers. Often in industry the work station is designed in an arbitrary manner, giving little consideration to the anthropometric measurements of the anticipated user. This situation is aggravated by the unavailable design parameters or dimensions.

An ergonomics approach to industrial work station design attempts to achieve an appropriate matching of worker capabilities and the work requirements to optimize productivity as well as to provide for worker physical and mental well-being, job satisfaction, and safety. Many theories, principles, methods, and data relevant to such design objectives have been generated through ergonomics research. This chapter will present sets of generally accepted design parameters and guidelines for industrial work station design.

EVOLUTION OF INDUSTRIAL WORK STATION DESIGN

Industrial work station design concepts have evolved mainly from industrial engineering, industrial hygiene, safety engineering, industrial relations, behavioral science, industrial psychology, and ergonomics. These disciplines have one or more of the following objectives for industrial work stations:

- Reduce worker fatigue
- Improve working environments
- Minimize worker safety hazards

Industrial engineering focuses on plant layout, work methods, and work measurement. Plant layout deals with overall facilities and with the arrangement of the individual work areas. By pursuing the principles of motion economy, the designer can improve worker productivity and reduce fatigue in the performance of a manual task. Table 11-1 presents the principles of motion economy with regard to the arrangement of the work station.[5, 6] In the systematic procedure of developing new work stations and improving methods at existing work centers, a necessary step is determining time standards through appropriate work measurement techniques.

Industrial hygiene and safety engineering ensure that the environment and work station do not pose adverse health and safety hazards to workers. The Occupational Safety and Health Administration (OSHA) has made a major impact on setting safety standards and enforcing them at the work station through periodic inspection process. Behavioral science, industrial relations, and industrial psychology are all interested in improving work organization and job design to enhance worker motivation, satisfaction, and job attitudes.

Table 11-1. Principles of Motion Economy for the Arrangement of the Work Station.

1.	Provide definite and fixed stations for all tools and materials to permit habit formation.
2.	Position tools and materials to reduce searching.
3.	Use gravity feed, bins and containers to deliver the materials as close to the point of use as possible.
4.	Locate tools, materials and controls within the normal working area if possible and if not within the maximum working area.
5.	Arrange materials and tools to permit the best sequence of motions.
6.	Use "drop deliveries" or ejectors wherever possible so that the operator does not have to use his or hands to dispose of the finished work.
7.	Provide adequate lighting, and a chair of the type and height to permit good posture. Arrange the height of the work station to allow alternate standing and sitting.
8.	Contrast the color of the work station with that of the work to reduce eye fatigue.

Ergonomics is concerned with biomechanics, kinesiology, work physiology, and anthropometry. *Biomechanics* is the mechanics of biologic systems, especially the human body. The biomechanical approach to industrial work station design primarily considers worker's capabilities, task demands, and equipment in an integrated manner. *Kinesiology* deals with the study of human movement in terms of functional anatomy. The principles of kinesiology should be used in the design of work stations to avoid incompatible movements.

Work physiology determines workers' physiologic reactions to their job demands and maintains the workers' physiologic cost within safe limits. Workers vary in their age, sex, background, physical, and mental characteristics and health. These factors must be considered in the design of work stations to maintain high productivity over a time period, during which different workers will perform the same task.

Anthropometry is mainly concerned with work station dimensions and the arrangement of tools, equipment, and material. The anthropometric data consist of worker body dimensions, ranges of motions for arms and legs, and muscle strength capabilities. Designers should take advantage of the available anthropometric data to determine work station dimensions.[7] The discipline is considered vital in the development of industrial workstation design parameters.

INDUSTRIAL WORK STATION DESIGN PRINCIPLES

From an anthropometric point of view, Sanders and McCormick[8] suggest the following general approach to design an industrial work station:

- Determine the body dimensions that are relevant to the design, such as sitting height
- Define the user population who is expected to use the equipment or facilities
- Determine the user criteria, such as design for the average individual, the extreme individual, or for an adjustable rang
- Select the relevant percentage of the population to be accommodated by the design, such as 90%, 95%, or whatever is appropriate to the problem
- Locate appropriate anthropometric tables and extract relevant values
- Add appropriate allowances, if special clothing is to be worn

Additional principles have been proposed over the years. Woodson and Conover[9] recommend that the dimensions of larger workers should be used for determining clearances, while those of smaller workers should be used to determine limits of reach. They also point out that clothing can add to the clearance requirements and may restrict movement.

Khalil[10] contends that for a successful industrial work station design, the designer should follow four basic design rules:

- Recognize that the worker is the center of the design, consider the worker's anatomic structure, and obtain accurate anthropometric dimensions to fit the job to the worker.
- Use kinesiology principles in the design and avoid incompatible movements.
- Observe the worker's physiologic capacity and use physiologic responses as the criteria for the design.
- Apply psychological principles for improving morale and satisfaction.

The previously developed guidelines for an industrial work station design are limited in application, because they neglect necessary anthropometric considerations due to different ethnic compositions of the diverse working populations.[10] To obtain maximum worker efficiency and physical well-being, Tichauer[1] suggests seven guidelines for determining work station dimensions (Table 11-2). Konz[11] recommends 16 principles concerning physical design of the industrial work stations (Table 11-3).

A serious error when designing a work station is to only fit the average person, because no one is average in all aspects. This is known as the "fallacy of the average person." Hertzberg[4] recommends that the designer pay special attention to what the worker must see, hear, reach, and manipulate as well as to the body clearance.

Table 11-2. Guidelines for Determining Work Station Dimensions.

1. Consider body measurement as well as range and strength of movements of the kinetic elements involved in a task.
2. Take account of the full range of body measurements of the specific working population involved in a task.
3. Recognize the differences of sex, ethnic groups, and educational or social background because they affect specific types of manipulative skills.
4. Determine by biomechanical analysis the optimal position of an operator with regard to equipment controls.
5. Provide maximum postural freedom in task design, especially for repetitive work situations.
6. Avoid standing on the toes and torsion or sideways bending of the trunk.
7. Refrain from performing a task on a strip 3 inches (7.6 cm) wide from the border of the bench which is closest to the operator.

Work Station Height

The working height of the work station is also important. When the work area is too high, workers must frequently lift their shoulders up, and this may cause painful cramps near the shoulder blades and in the neck and shoulders. On the other hand, when the working height is too low, the back must be excessively bowed and this causes backache. Consequently, the work table height must be compatible with worker height, whether standing or sitting. To determine proper work station height, Konz[11] recommends three different approaches:

- Make available several work station or work surface heights
- Adjust the worker's elbow rather than the height of the work station
- Adjust the work on the workbench

Tool Location

A primary requirement for assembly and production operations is that all materials, tools, and equipment, which require manual operation, be located so that all workers can effectively reach and operate them. The manner in which the human-machine interface is accomplished in an industrial work station design has important implications for the system's performance as well as for the workers' well-being and safety.[12] The workers interact with the machine or equipment through control devices such as knobs, levers, and pedals, and displays such as dials, counters, and lights. The arrangement of components in a work station design is made on the basis of the following principles:

- Operational importance
- Frequency of use
- Function
- Sequence of use

Table 11-3. Physical Design Principles of the Industrial Work Station.

1.	Avoid static loads and fixed work postures.
2.	Set the work height at 2 in below the elbow.
3.	Furnish every operator with an adjustable chair.
4.	Support limbs.
5.	Use the feet as well as the hands.
6.	Use gravity, do not oppose it.
7.	Conserve momentum.
8.	Use two-handed motions rather than one-handed motions.
9.	Use rowing motions for eye control for two-handed motions.
10.	Use rowing motions for two-handed motions.
11.	Pivot movements about the elbow.
12.	Use the preferred hand.
13.	Keep arm motions in the normal working area.
14.	Let the small woman reach; let the large man fit.
15.	Locate all materials, tools, and controls in a fixed place.
16.	View large objects for a long time.

The principles of operational importance and frequency of use are especially applicable for locating components in a general area, whereas the sequence of use and functional principles seem to apply more to the arrangement of components within a general area. Little empiric evidence is available regarding the evaluation of these principles.[8] Pulat and Ayoub[13] have developed a computerized algorithm, which can be used to select displays and controls for process control jobs. The following lists some general guidelines for the arrangement of displays and controls:[14]

- First priority - Primary visual tasks
- Second priority - Primary controls that impact with primary visual tasks
- Third priority - Control-display relationships
- Fourth priority - Arrangement of elements to be used in sequence
- Fifth priority - Convenient location of frequently used elements
- Sixth priority - Consistency with other layouts within the system or in other systems

For a seated worker at a fixed workstation, Van Cott and Kinkade[14] suggest a minimum workspace measure of 30.0 in x 30.0 in x 57.6 in. The controls, switches, tools, equipment, and other objects to be handled should be placed within easy reach and in the most appropriate spatial position relative to the worker.

Industrial work space design dimensions must be compatible with the anthropometric considerations of the anticipated user, which will result in greater productivity by maximizing efficiency at a minimum human cost. Unfortunately, poorly designed work stations and work spaces are still common in industry due to lack of usable design information.

Maynard[15] presented the concept of normal and maximum working areas in the horizontal and vertical planes to improve the design of work space. Maynard did not provide any dimensions to the work spaces, so this concept offers minimal guidance to a designer. Barnes[16] gave dimensions to Maynard's work spaces, based on the data generated by Asa.[17]

Farley[18] also gave new dimensions to the work spaces proposed by Maynard (Figures 11-1 and 11-2), including only one work area for men and one for women, based on their average physical dimensions. He determined the normal working area as being equal to the volume circumscribed by the horizontal arm pivoting about a relaxed vertical arm. The volume circumscribed during the movement of the fully extended arm pivoting about the shoulder pivot point represented the maximum working area. The results reported by Farley are presently the most widely used in industry.

Squires[19] pointed out that the elbow does not stay at a fixed point, but moves out and away from the body as the forearm pivots. He determined only the normal horizontal working area based on the existing anthropometric data for the 10th percentile male (Figure 11-3). Squires' concept provided a substantial enhancement over Farley's concept in the one-handed, two-handed, and total normal front range for all types of workers and resulted in a relatively small decrease in the normal side range in all but one instance. Considering the overall increase in the normal working area and the dynamic nature of the arm movement while the hand sweeps an arc in the normal horizontal working area, Squires' concept is preferred to Farley's concept.[20]

Many industrial operations require workers to perform their tasks while sitting, standing, or in a combination of sitting and standing positions. In industry, a particular work station may be used by one or several men, one or several women, or a combination of men and women. It is not always feasible or practical to change the design parameters every day or each work shift to match the individual worker's anthropometric characteristics. The best designed industrial work station is flexible and accommodates the range of anthropometric characteristics of the workers who will use it.

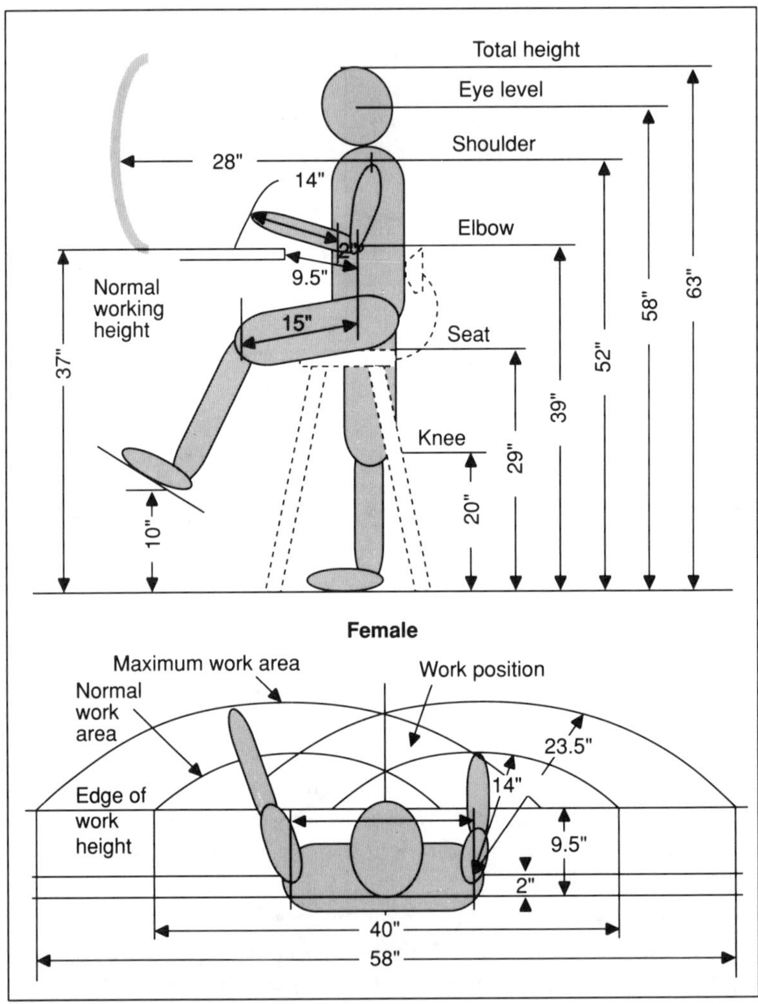

Figure 11-1. Normal and Maximum Working Areas in the Horizontal and Vertical Planes for Female Operator.[18]

Note: Conversion factor for SI Units: 1 in = 2.54 cm.

APPLYING ENGINEERING ANTHROPOMETRY TO INDUSTRIAL WORK STATION DESIGN

Das and Grady[20, 21, 23] determined design dimensions for industrial work stations through the use of the existing anthropometric data, so that these dimensions can be readily employed by a designer. Work space design dimensions were

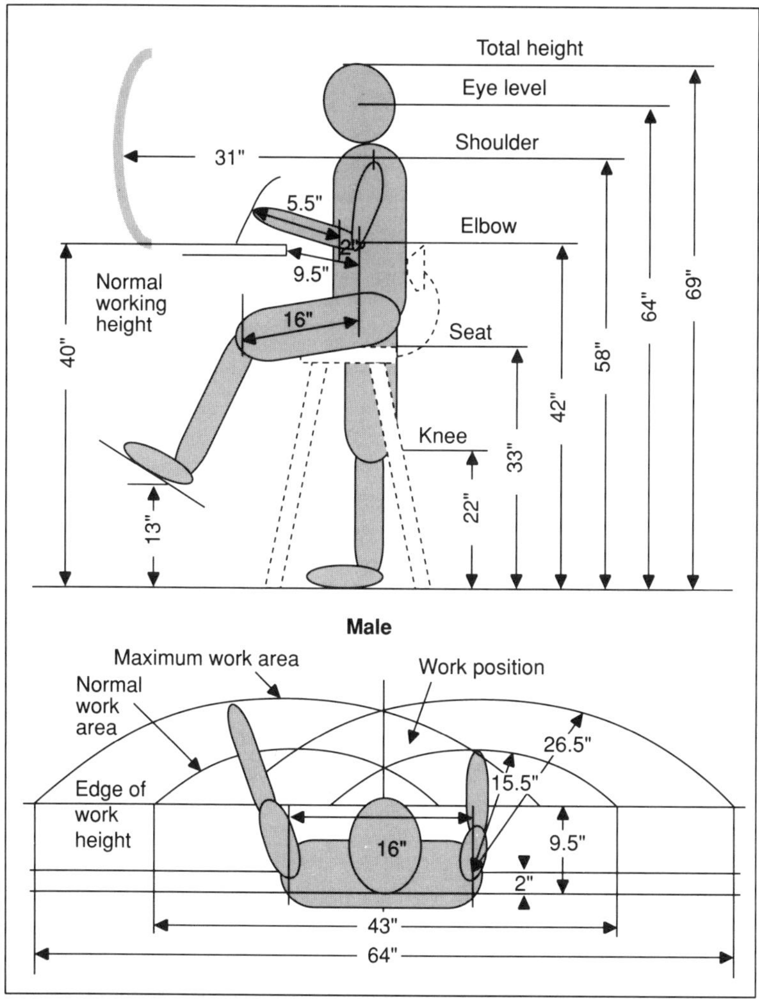

Figure 11-2. Normal and Maximum Working Areas in the Horizontal and Vertical Planes for Male Operator.[18]
Note: Conversion factor for SI Units: 1 in = 2.54 cm.

determined for industrial tasks in sitting, standing, and sit-stand positions. Worker populations consisted of a combination of male and female workers and the individual male and female workers for the 5th, 50th, and 95th percentiles, based on existing anthropometric data for the US military population.[24] The source data are often used in the United States to design work space, equipment, and clothing. In the future using dimensions of a civilian or industrial population to develop industrial work space design dimensions will be more desirable. Currently, such an endeavor is in progress.[25]

The normal and maximum reach dimensions were based on the most commonly used industrial operations, which require a grasping movement or thumb and forefinger manipulation. Appropriate allowances were provided to adjust the reach dimensions for other types of industrial operations, however. The normal and maximum horizontal and vertical clearances and reference points for the horizontal and vertical clearances were established to facilitate the design. The concepts developed by Farley[18] and Squires[19] were used to describe the work space envelope for the individual worker. The dimensions of smaller (5th percentile) and larger (95th percentile) workers were used for determining the limits of reach and clearance requirements, respectively.[9]

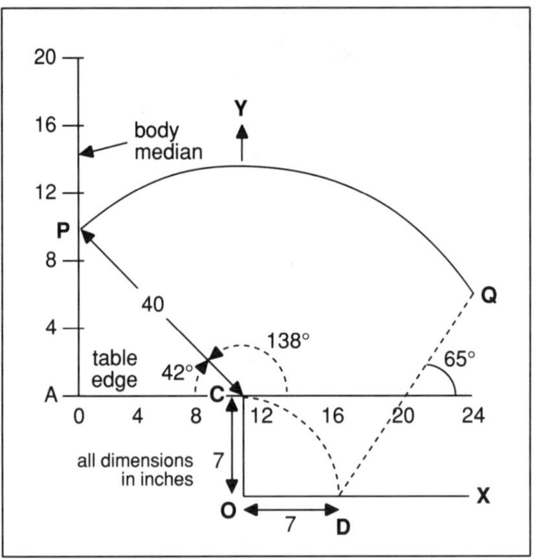

Figure 11-3. Normal Working Areas in the Horizontal Plane for Male Operator.[19]
Note: Conversion factor for SI Units: 1 in = 2.54 cm.

Adjustment of Existing Anthropometric Data

The existing anthropometric data were based on erect standing and sitting positions. The industrial worker seldom assumes erect positions at work or at rest. Consequently, the data were adjusted to account for the "slump" posture involved in the "normal" standing and sitting positions (Table 11-4). Furthermore, the existing anthropometric data were derived on the basis of measurements from nude subjects.[4] The data, therefore, were duly adjusted for clothing, shoe, and other allowances (Table 11-4). Table 11-5 presents these adjusted anthropometric measurements, which have been used to generate the pertinent design parameters of work space.

Table 11-4. Clothing, Shoe, and Other Allowances or Adjustments.

Item	Body Feature	Allowances or Adjustments*
A	Total height	Add 1.0 in for men's and women's shoes. Subtract 0.75 in for slump posture.
B	Body height (sitting)	Add 0.25 in for clothing under the buttocks. Subtract 1.75 in for slump posture.
C	Eye height	Add 1.0 in for men's and women's shoes. Subtract 0.75 in for slump posture.
D	Eye height (sitting)	Add 0.25 in for clothing under the buttocks. Subtract 1.75 in for slump posture.
E	Shoulder height	Add 1.0 in for men's and women's shoes and 0.55 in for clothing.
F	Shoulder height (sitting)	Add 0.25 in for clothing under the buttocks.
G	Body depth	Add 0.40 in for clothing.
H	Elbow-to-elbow	Add 0.55 in for clothing.
I	Thigh clearance	Add 0.80 in for clothing.
J	Forearm length	Subtract 3.0 in for thumb and forefinger manipulation.
K	Arm length	Subtract 2.0 in for measurement from the back of the shoulder. Subtract 3.0 in for thumb and forefinger manipulation.
L	Elbow height	Add 1.0 in for men's and women's shoes.
M	Elbow height (sitting)	No allowance.
N	Knee height (sitting)	Add 1.0 in for men's and women's shoes.

* Conversion factor to SI Unit: 1 in = 2.54 cm.

INDUSTRIAL WORK SPACE DIMENSIONS

For the Seated Worker

Preferably an industrial operation should be performed in a seated position to improve worker productivity by maximizing effective motions, reducing worker fatigue, and increasing worker stability and equilibrium. A seated position has a number of advantages over a standing position, such as lower physiologic load when sitting, reduced muscular load required to maintain body posture, and improved blood circulation.

Prolonged sitting may cause pain, however, especially in the spine and in the muscles of the back. The situation is aggravated when poorly designed chairs are used. Prolonged sitting may also lead to curvature of the spine and slackening of the abdominal muscles. Kvalseth[12] maintains that seated operation is preferable for jobs that require:

- Fine manipulative hand movements
- A high degree of body stability and equilibrium
- Precise foot control action
- When all materials and tools are located within the seated work space
- Heavy material handling
- Fixed body posture for extended time periods

Table 11-5. Corrected Anthropometric Measurements to Account for Shoe, Clothing, and Other Allowances or Adjustments.

			Percentiles (inches)		
Sex	Item	Body Feature	5th	50th	95th
Male	A	Total height (slump)	65.5	69.4	73.4
	B	Body height (sitting, slump)	32.3	34.5	36.5
	C	Eye height (slump)	61.1	65.0	68.9
	D	Eye height (sitting, slump)	27.9	30.0	32.0
	E	Shoulder height	54.4	58.2	61.8
	F	Shoulder height (sitting)	21.6	23.6	25.4
	G	Body depth	10.5	11.9	13.4
	H	Elbow-to-elbow	15.8	17.8	20.4
	I	Thigh clearance	5.6	6.4	7.3
	J	Forearm length	14.6	15.9	17.2
	K	Arm length	26.9	29.6	32.3
	L	Elbow height	41.6	44.5	47.4
	M	Elbow height (sitting)	7.4	9.1	10.8
	N	Popliteal height (sitting)	16.7	18.0	19.2
Female	A	Total height (slump)	60.6	64.3	68.5
	B	Body height (sitting, slump)	30.9	32.6	34.3
	C	Eye height (slump)	56.9	60.5	64.3
	D	Eye height (sitting, slump)	27.3	28.8	30.4
	E	Shoulder height	50.4	54.2	57.8
	F	Shoulder height (sitting)	20.6	22.6	24.4
	G	Body depth	8.6	9.7	10.9
	H	Elbow-to-elbow	14.1	15.5	17.3
	I	Thigh clearance	4.9	5.7	6.5
	J	Forearm length	12.8	14.4	16.0
	K	Arm length	23.7	26.0	28.5
	L	Elbow height	39.0	41.4	43.8
	M	Elbow height (sitting)	7.4	9.1	10.8
	N	Popliteal height (sitting)	14.7	16.0	17.2

Dimensions based on the US Military population. Conversion factor to SI Unit: 1 in = 2.54 cm.

In an industrial work station an adjustable height chair is very desirable because a standard workbench height, which of course is dependent on the nature of the work, can be used in the design. The work height should be determined from elbow height, not distance from the floor, and the chair should be adjusted so that the optimum work height can be maintained. The recommended workbench height is about 2 in below the elbow.[14]

Figure 11-4 shows the horizontal normal and maximum working areas and clearances for a seated or standing male or female worker. The normal and maximum working areas in both the horizontal and vertical plane were projected on the shoulder pivot point (P). The pivot point in the horizontal plane was located one-half of the body depth from where the worker's back made contact with the chair back. The 95th percentile body depth (G) measurement was used to locate the chair back. The length (S) represented the distance from the shoulder pivot point to the workbench edge in both the horizontal and vertical planes.

In Figure 11-5 the shoulder pivot point in the vertical plane was located from the chair seat. The 95th percentile thigh clearance (I) was used to locate the chair seat. Also, the elbow height (M), the popliteal height, sitting (N) and the body height (B) were based on the 95th percentile measurements. The vertical location of the shoulder pivot point was determined so that smaller workers could reach and properly view their work; the shoulder height (F) and the eye level (D) were based on the 5th percentile measurement. The workbench or elbow height (M) was based on the 95th percentile value.

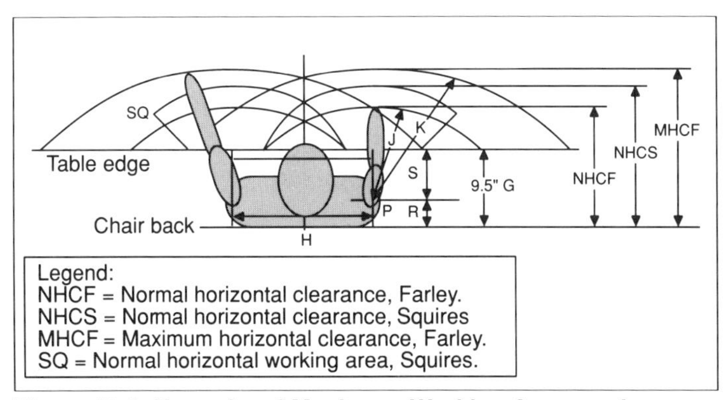

Figure 11-4. Normal and Maximum Working Areas and Clearances in the Horizontal Plane for a Seated (also for standing and sit-stand) Male or Female Operator (dimensions shown in Tables 11-6 and 11-7).

The normal and maximum working areas in the horizontal and vertical planes were determined by employing Farley's concept.[18] The fifth percentile values of the forearm length (H), the arm reach (K), and the elbow-to-elbow breadth (H) were used for this purpose. To apply Squires' concept of the normal

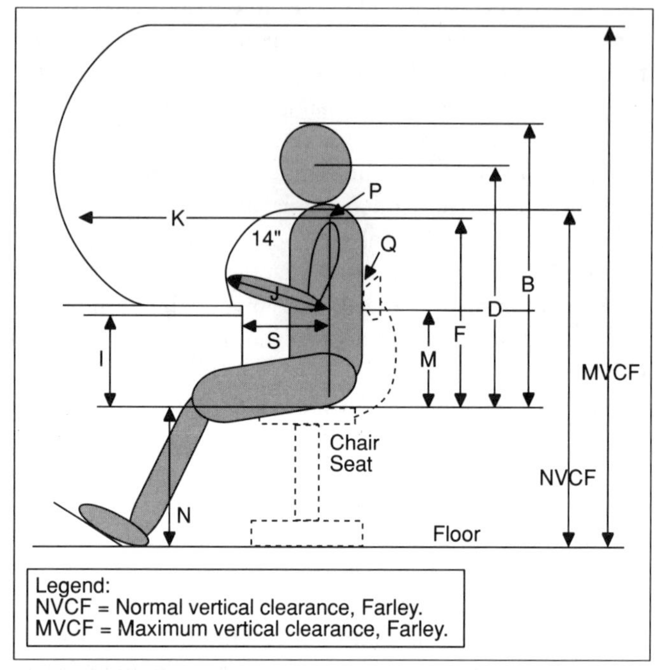

Figure 11-5. Normal and Maximum Working Areas and Clearances in the Vertical Plane for a Seated Male or Female Operator (dimensions shown in Tables 11-6 and 11-7).

working area in the horizontal plane,[19] it was necessary to use the distance between the elbow and the shoulder projection (S). Table 11-6 summarizes the pertinent design dimensions for workspace for the seated worker. The normal and maximum horizontal and vertical clearance dimensions for workspace for seated operators are summarized in Table 11-7.

For the Standing Worker

Standing work positions are less desirable, but are often necessary in industry. When workers are required to stand for extended periods of time, blood and other fluids tend to accumulate in the legs and result in venous pooling, swelling of the legs, and varicose veins. Movement of the legs helps move the blood and prevent these problems. Kvalseth[12] contends that standing operations are preferable for jobs that require:

- Frequent handling of heavy objects
- Extended reaches and moves of substantial magnitude and frequency
- Mobility to move around the workstation
- Manual downward forces of substantial magnitude and frequency

Table 11-6. Dimensions for Workplace Layouts for Seated Male and Female Operators.

Item	Body Feature	General Male (inches)	General Female (inches)	Combination of Male and Female (inches)	Individual Male, Percentiles (inches)			Individual Female, Percentiles (inches)		
					5th	50th	95th	5th	50th	95th
B	Body height (sitting, slump)	36.5	34.3	36.5	32.3	34.5	36.5	30.9	32.6	34.3
D	Eye height (sitting, slump)	27.9	27.3	27.3	27.9	30.0	32.0	27.3	28.8	30.4
F	Shoulder height (sitting)	21.6	20.6	20.6	21.6	23.6	25.4	20.6	22.6	24.4
G	Body depth	13.4	10.9	13.4	10.5	11.9	13.4	8.6	9.7	10.9
H	Elbow-to-elbow	15.8	14.1	14.1	15.8	17.8	20.4	14.1	15.5	17.3
I	Thigh clearance	7.3	6.5	7.3	5.6	6.4	7.3	4.9	5.7	6.5
J	Forearm length	14.6	12.8	12.8	14.6	15.9	17.2	12.8	14.4	16.0
K	Arm reach	26.9	23.7	23.7	26.9	29.6	32.3	23.7	26.0	28.5
M	Elbow height (sitting)	10.8	10.8	10.8	7.4	9.1	10.8	7.4	9.1	10.8
N	Popliteal height (sitting)	19.2	17.2	19.2	16.7	18.0	19.2	14.7	16.0	17.2

1. (i) All dimensions based on US military population; (ii) normal and maximum reach based on thumb and forefinger manipulation (J and K); (iii) adjust J and K values: end-of-fingertip add 3 inches; push add 2 inches; flip add 2.5 inches; and grip subtract 2 inches.
2. Conversion factor to SI Unit: 1 in = 2.54 cm.

For a standing operation, the designer should give special attention to the height of the workbench. An adjustable workbench is ideal since the entire workplace population can be accommodated. In an industrial situation, however, the same workbench may be used by a variety of workers, which could result in frequent adjustment of the workbench height. As an alternate solution, the platform on which the worker stands could be raised or lowered to adjust the worker's elbow height. The platform can be as simple as several pieces of rug piled on top of each other. Rugs help reduce fatigue to the feet and back.[11]

The horizontal working area for a standing male or female worker is shown in Figure 11-4. The distance of the shoulder pivot point (P) in the horizontal plane from the workbench edge is one-half of the body depth (G).

Figure 11-6 shows the vertical working area for a male or female worker. The location of point P in the vertical plane would be established from the floor. The distance from the floor to the shoulder height would be based on the fifth

Table 11-7. Normal and Maximum Horizontal and Vertical Clearance Dimensions for Workplace Layouts for Seated Male and Female Operators.

Population	Horizontal Clearance (inches)			Vertical Clearance (inches)	
	Normal Farley (NHCF)	Normal Squires (NHCF)	Maximum Farley (MHCF)	Normal Farley (NVCF)	Maximum Farley (MVCF)
General male	21.3	33.6	28.0	44.6	67.7
General female	18.3	29.2	23.7	40.8	61.5
Combination of male and female	19.5	30.4	26.2	42.8	63.5
5th percentile male	19.8	32.1	25.1	38.7	65.2
50th percentile male	21.8	35.5	27.8	43.0	71.2
95th percentile male	23.9	39.0	30.6	47.2	76.9
5th percentile female	17.1	28.0	21.4	34.9	59.0
50th percentile female	19.2	30.8	24.1	39.5	64.6
95th percentile female	21.4	33.9	26.9	44.0	70.1

1. (i) Reference points for horizontal and vertical clearances are chair back and floor, respectively; and (ii) normal and maximum reach based on thumb and forefinger manipulation.
2. Coversion factor to SI Unit: 1 in = 2.54 cm.

percentile shoulder height (E) to facilitate the reach of the smaller workers. Since a platform would be provided to adjust the elbow height of the smaller workers and removed for the larger workers, the dimensions of the body height (A) and the elbow height (L) would be based on the fifth percentile measurements. The eye level (C) would be based on the fifth percentile measurement, so that the smaller workers could see their work properly.

The normal and maximum working areas in the horizontal and vertical planes would be determined in the same manner as described for the seated worker. Table 11-8 summarizes the pertinent design dimensions for work space for the standing worker. Table 11-9 summarizes the normal and maximum horizontal and vertical clearance dimensions for work space for the standing worker.

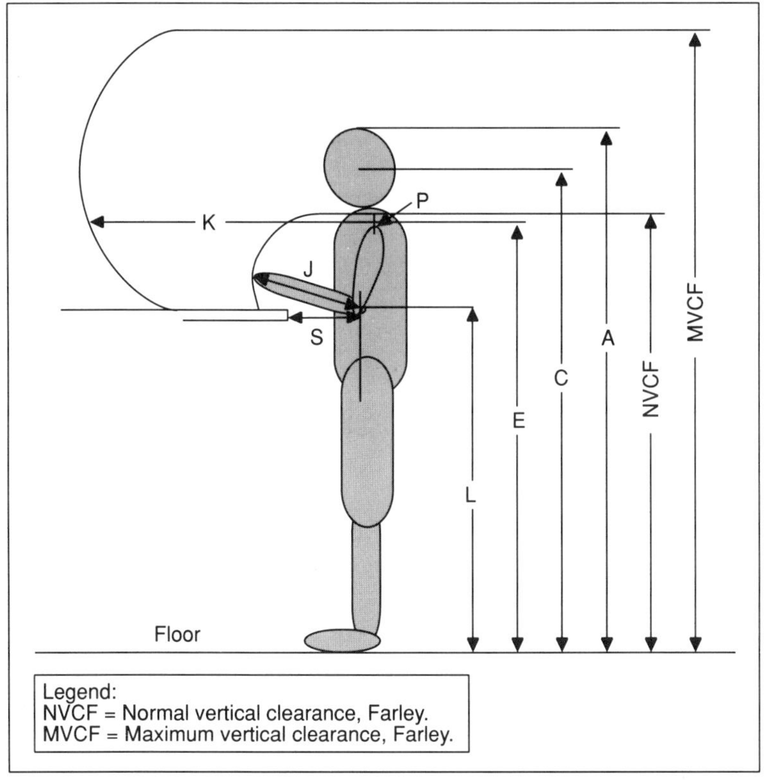

Figure 11-6. Normal and Maximum Working Areas and Clearances in the Vertical Plane for a Standing Male or Female Operator (dimensions shown in Tables 11-8 and 11-9).

For the Sit-Stand Worker

Generally a sit-stand work position is more desirable than only a sitting or only a standing position, because such a work position permits workers to shift their postures at will and reduce the muscular fatigue, which results from a prolonged effort in one position. A work station, which permits workers to sit or stand as they wish, is highly desirable from a physiologic and orthopedic viewpoint. Standing and sitting impose stresses on different muscle groups, consequently each changeover relaxes some muscles and stresses others.

When designing a sit-stand operation, the following recommendations should be observed:

- Provide a raised seat, since the worker alternates between the sitting and standing positions
- Ensure the seat is easily moveable into and out of position
- Provide a footrest for the seated position to minimize muscle soreness and fatigue.

Table 11-8. Dimensions for Workplace Layouts for Standing Operators.

Item	Body Feature	General Male (inches)	General Female (inches)	Combination of Male and Female (inches)	Individual Male, Percentiles (inches)			Individual Female, Percentiles (inches)		
					5th	50th	95th	5th	50th	95th
A	Total height (slump)	73.4	68.5	73.4	65.5	69.4	73.4	60.6	64.3	68.5
C	Eye height (slump)	61.1	56.9	56.9	61.1	65.0	68.9	56.9	60.5	64.3
E	Shoulder height	54.4	50.4	50.4	54.4	58.2	61.8	50.4	54.2	57.8
G	Body depth	13.4	10.9	13.4	10.5	11.9	13.4	8.6	9.7	10.9
H	Elbow-to-elbow	15.8	14.1	14.1	15.8	17.8	20.4	14.1	15.5	17.3
J	Forearm length	14.6	12.8	12.8	14.6	15.9	17.2	12.8	14.4	16.0
K	Arm reach	26.9	23.7	23.7	26.9	29.6	32.3	23.7	26.0	28.5
L	Elbow height	47.4	43.8	47.4	41.6	44.5	47.4	39.0	41.4	43.8

1. (i) All dimensions based on US military population; (ii) normal and maximum reach based on thumb and forefinger manipulations (J and K); and III) adjust J and K values for: end of fingertip add 3 inches; push add 2 inches; flip add 2.5 inches; and grip subtract 2 inches.
2. Conversion factor to SI Unit: 1 in = 2.54 cm.

Work space dimensions for this work station should include design parameters from sitting and standing work stations. The working area in the horizontal and vertical planes for the sit-stand male or female worker are shown in Figures 11-4 and 11-7 respectively.

To determine the location of the shoulder pivot point (P) in the horizontal plane with respect to the workbench edge, the same procedure as outlined for the seated worker would be used. The point P in the vertical plane is located in the same manner as described for the standing worker. The body height (A), the thigh clearance (I), the elbow height (L) and the popliteal height, and sitting (N) were based on the 95th percentile measurements, and the eye level (C) would be based on the fifth percentile measurement.

The normal and maximum working areas in the horizontal and vertical planes could be determined by following the same procedures as outlined earlier. Table 11-11 summarizes the design dimensions for sit-stand work spaces. In addition, Table 11-4 summarizes the normal and maximum horizontal and vertical clearance dimensions for sit-stand (also for standing) workers.

Das[26] has shown the manner by which engineering anthropometry can be used effectively for the work space design of a power-feed drill-press operation, with jig and fixture. In this operation, workers producted a connector plate by

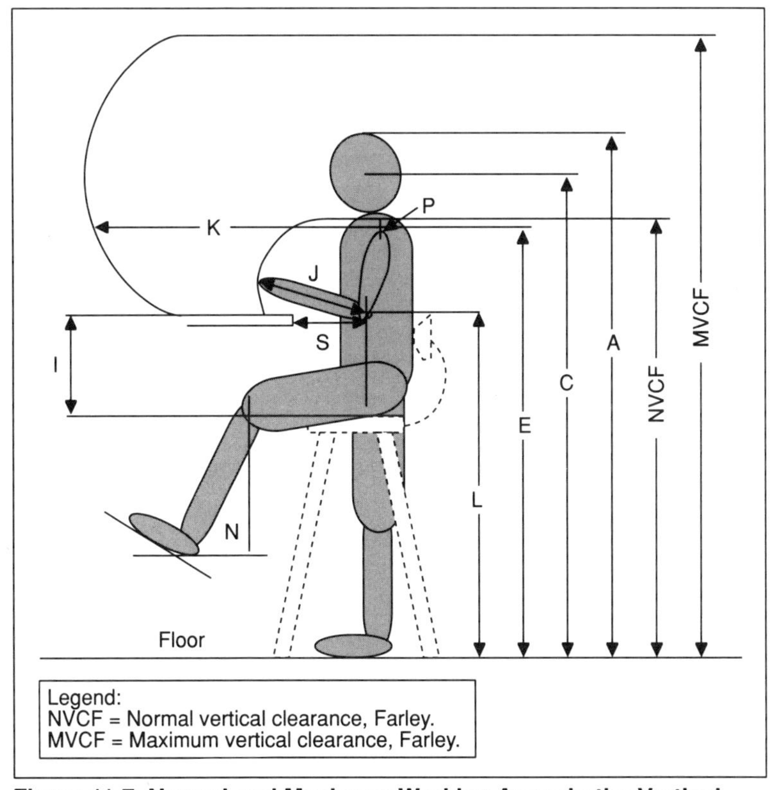

Figure 11-7. Normal and Maximum Working Areas in the Vertical Plane for a Sit-stand Male or Female Operator (dimensions shown in Tables 11-9 and 11-10).

drilling four holes on a prepared steel plate. The task was performed in a seated position by both male and female workers (Figures 11-8 and 11-9).

COMPUTER AIDED WORK SPACE DESIGN

The modeling of the human body can be visualized as a link and joint system, which has its analogue in mechanical designs. Computerized human movement models and computer-generated work space elements help designers analyze human movements through the use of simulated subjects or simulated limbs on a computer display.[27]

Dynamic models, as opposed to static models, are not intended to reproduce anthropometric data for work space design. The dynamic mechanical and biologic models deal with the human body's mass distribution, muscle forces, and tissue influences.

Computer simulations have been developed to simulate three-dimensional space, an anthropometric variable human model in a realistic work space geometry. Initially the computer human model was started with the development of stickman or linkman models. To simulate link positions, the models

Table 11-9. Normal and Maximum Horizontal and Vertical Clearance Dimensions for Workplace Layouts for Standing (also for sit-stand) Operators.

Population	Horizontal Clearance (inches)			Vertical Clearance (inches)	
	Normal Farley (NHCF)	Normal Squires (NHCF)	Maximum Farley (MHCF)	Normal Farley (NVCF)	Maximum Farley (MVCF)
General male	21.3	33.6	28.0	62.0	81.3
General female	18.2	29.1	23.7	56.6	74.1
Combination of male and female	19.5	30.4	26.2	60.2	74.1
5th percentile male	19.8	32.1	25.1	56.2	81.3
50th percentile male	21.8	35.5	27.8	60.4	87.8
95th percential male	23.9	39.0	30.6	64.6	94.1
5th percentile female	17.1	28.0	21.4	40.8	74.1
50th percentile female	19.2	30.8	24.1	44.5	80.2
95th percentile female	21.4	33.9	26.9	48.8	86.3

1. (i) Reference points for horizontal and vertical clearances are operator back and floor, respectively, and (ii) normal and maximum reach based on thumb and forefinger manipulation.
2. Conversion factor to SI Unit: 1 in = 2.54 cm.

used anthropometric databases. Later more sophisticated and functional analogues such as BOEMAN (developed for the Boeing Company), COMBIMAN (Computerized Biomechanical Man-Model), and SAMMIE (Systems of Aiding Man-Machine Interaction Evaluation) were developed.

The BOEMAN model produces a projection of a three-dimensional variable geometric mode (Figure 11-10).[28] The COMBIMAN is a three-dimensional variable geometric mode (Figure 11-11)[29], and is constructed initially of 33 links that correspond functionally to the human skeletal system. The SAMMIE computer-aided design system allows the designer to view three-dimensional work stations on the screen of a graphics terminal (Figure 11-12).[30] A man model of variable anthropometry can be added to the work space model, enabling the work space to be evaluated against such criteria as a worker's reach, vision, and fit capabilities.

References

1. Tichauer ER. *Occupational Biomechanics: The Anatomical Bases of Workplace Design.* New York, NY: Institute of Rehabilitation Medicine; 1975; New York University Medical Centre Rehabilitation Monograph No. 51.

2. Corlett EN, Bowssenna M and Pheasant ST. Is discomfort related to the postural loading of the joints? *Ergonomics.* 1982;25:315-322.

3. Grandjean E, Hunting W, Maeda K and Laudi T. Constrained postures at office workstations. In: Kavelseth TO, ed. *Ergonomics of Workstation Design.* Kent, England: Butterworth Publishers; 1983.

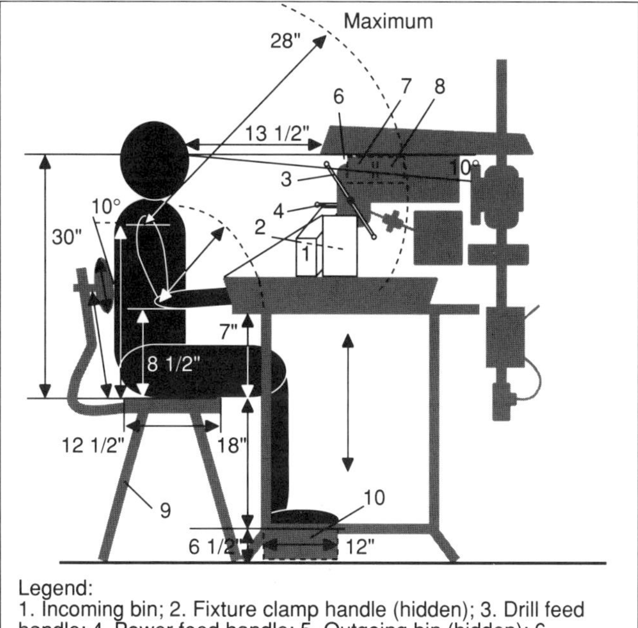

Legend:
1. Incoming bin; 2. Fixture clamp handle (hidden); 3. Drill feed handle; 4. Power feed handle; 5. Outgoing bin (hidden); 6. Electric totalizing counter (hidden); 7. Production standard stand; 8. Quality feedback stand; 9. Chair; 10. Foot rest.
Note: Normal and maximum working areas in the vertical plane, based on women. Conversion factor to SI Units: 1 in = 2.54 cm.

Figure 11-8. Workspace Design for Drill Press Operation - Side View.

4. Hertzberg HTE. Engineering anthropology. In: Van Cott HP and Kinkade RG, eds. *Human Engineering Guide to Equipment Design, Revised Edition*. New York, NY: McGraw-Hill; 1972.

5. International Labour Office. *Introduction to Work Study, Revised Edition*. Switzerland: Geneva; 1974.

6. Niebel BW. *Motion and Time Study*, 8th ed. Homewood, IL: Irwin; 1988.

7. Alexander DC. *The Practice and Management of Industrial Ergonomics*. Englewood Cliffs, NJ: Prentice-Hall; 1986.

8. Sanders MS and McCormick EJ. *Human Factors in Engineering and Design*. New York, NY: McGraw-Hill; 1987.

9. Woodson WE and Conover DW. *Human Engineering Guide for Equipment Designers*. Berkeley and Los Angeles, CA: University of California Press; 1964.

10. Khalil TM. Design tools and machines to fit the man. *Industrial Engineering*. Institute of Industrial Engineers; 1972;4:32-35.

11. Konz S. *Work Design: Industrial Ergonomics*, 2nd ed. Columbus, OH: Grid Publishing; 1983.

12. Kvalseth TO. Workstation design. In: Alexander DC and Pulat BM, eds. *Industrial Ergonomics: A Practitioners Guide*. Norcross, GA: Institute of Industrial Engineers; 1985.

13. Pulat BM and Ayoub MA. A computer aided display-control selection procedure

Table 11-10. Dimensions for Workplace Layouts for Sit-stand Male and Female Operators.

Item	Body Feature	General Male (inches)	General Female (inches)	Combination of Male and Female (inches)	Individual Male, Percentiles (inches)			Individual Female, Percentiles (inches)		
					5th	50th	95th	5th	50th	95th
A	Total height slump	73.4	68.5	73.4	65.5	69.4	73.4	60.6	64.3	68.5
C	Eye height (slump)	61.1	56.9	56.9	61.1	65.0	68.9	56.9	60.5	64.3
E	Shoulder height	54.4	50.4	50.4	54.4	58.2	61.8	50.4	54.2	57.8
G	Body depth	13.4	10.9	13.4	10.5	11.9	13.4	8.6	9.7	10.9
H	Elbow-to-elbow	15.8	14.1	14.1	15.8	17.8	20.4	14.1	15.5	17.3
I	Thigh clearance	7.3	6.5	7.3	5.6	6.4	7.3	4.9	5.7	6.5
J	Forearm length	14.6	12.8	12.8	14.6	15.9	17.2	12.8	14.4	16.0
K	Arm reach	26.9	23.7	23.7	26.9	29.6	32.3	23.7	26.0	28.5
L	Elbow height	47.4	43.8	47.4	41.6	44.5	47.4	39.0	41.4	43.8
N	Popliteal height (sitting)	24.3	21.8	24.3	21.1	22.7	24.3	19.1	20.5	21.8

1. (i) All dimensions based on US military population; (ii) normal and maximum reach based on thumb and forefinger manipulations (J and K); and (iii) adjust J and K values for: End of finger tip add 3 inches; push add 2 inches; flip add 2.5 inches; and grip subtract 2 inches.
2. Conversion factor to SI unit: 1 in = 2.54 cm.

for process control jobs - UNISER. *IIE Transactions.* Institute of Industrial Engineers; 1984;16(4):371-378.

14. Van Cott HP and Kinkade RG, eds. *Human Engineering Guide to Equipment Design.* New York, NY: McGraw-Hill; 1972.

15. Maynard HB. Workplace layouts that save time, effort and money. *Iron Age.* 1934;134:28-30;92.

16. Barnes RM. *Motion and Time Study*, 2nd ed. New York, NY: John Wiley; 1940.

17. Asa M. *A Study of Workplace Layout.* Ames, IA: State University of Iowa; 1942. M.Sc. Thesis.

18. Farley RR. Some principles of methods and motion study as used in development work. *General Motors Engineering Journal.* 1955;2:20-25.

19. Squires PE. *The Shape of the Normal Working Area.* New London, CT: U.S. Navy Department, Bureau of Medicine and Surgery; 1956; Report No. 256.

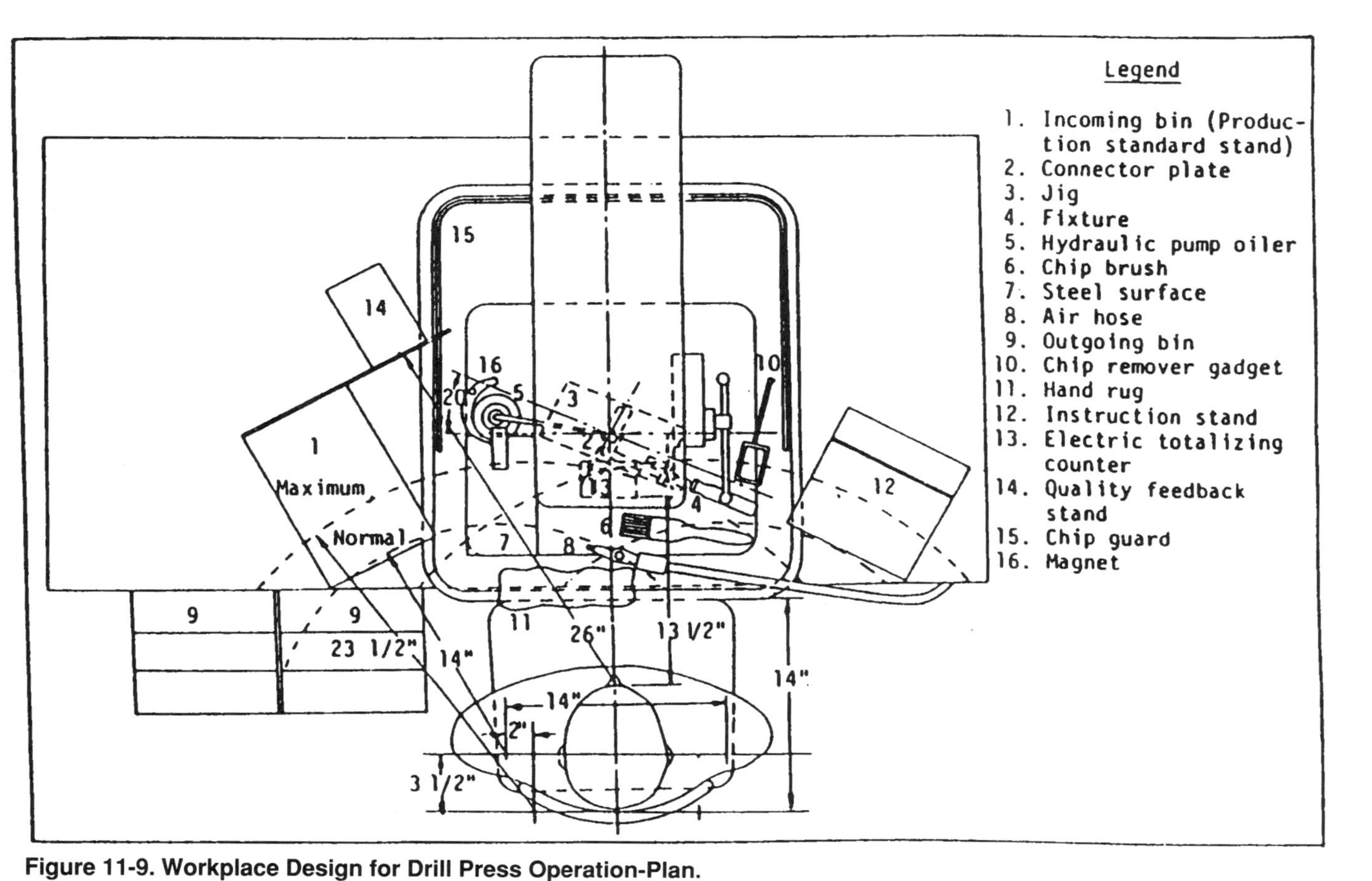

Figure 11-9. Workplace Design for Drill Press Operation-Plan.

Note: Normal and maximum working areas in horizontal plane based on women. Conversion factor to SI Unit: 1 in = 2.54 cm.

20. Das B and Grady RM. The normal working area in the horizontal plane: a comparative analysis between Farley's and Squires' concepts. *Ergonomics.* 1983b;26(5):449-459.

21. Das B and Grady RM. Industrial workplace layout design: an application of engineering anthropometry. *Ergonomics.* 1983a;26(5):433-447.

22. Konz S and Goel SC. The shape of the normal working area in the horizontal plane. *American Institute of Industrial Engineers Transactions.* 1969;1:70-73.

23. Das B and Grady RM. Industrial workplace layout and engineering anthropology. In: Kavelseth TO, ed. *Ergonomics of Workstation Design.* Kent, England: Butterworth Publishers; 1983c.

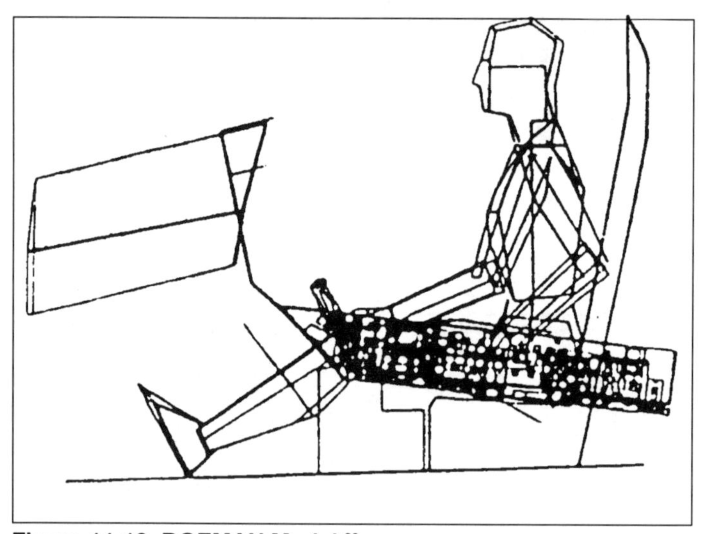

Figure 11-10. BOEMAN Model.[28]

24. Kennedy KW. *Reach Capability of the USAF Population.* Aerospace Medical Research Laboratory; 1964;TDR-64-69;AD 608269.

25. Das B and Behara DN. A new model for the determination of the horizontal normal working area. *Proceedings of the Annual International Industrial Ergonomics and Safety Conference*; 1989;195-202.

26. Das B. An ergonomics approach to the design of a manufacturing work system. *International Journal of Industrial Ergonomics.* 1987;1:231-240.

27. Holzhansen KD. Analysis of human movements for workplace design. In: Moraal J and Kraiss KF, eds. *Manned Systems Design: Methods, Equipment and Applications.* London: Plenum Press; 1981.

28. Ryan P. *Cockpit Geometry Evaluation, Phase II - Final Report, Vol. 1 - Program Description and Summary.* Seattle, WA: Boeing Company; 1971. JANAIR Report 7000201.

29. Kroemer KHE. *COMBIMAN*Computerized Biomechanical Man-Model.* Dayton, OH: Aerospace Medical Research Laboratory, Wright Patterson Air Force Base; 1972;AMRL-TR-72-16.

30. Bonney MC, Blunsdon CA, Case K and Porter JM. Man-machine interactions in work systems. *International Journal of Production Research.* 1979;17:619-624.

Biman Das, PhD, PE is Professor of Industrial Engineering, Technical University of Nova Scotia.

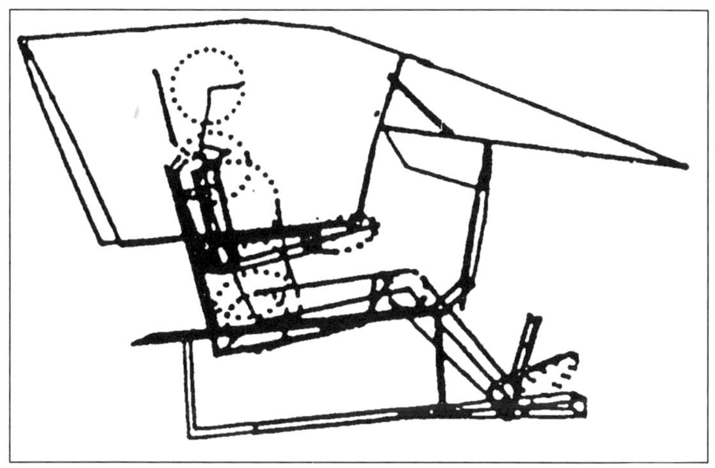

Figure 11-11. COMBIMAN Link-man in a Cockpit. [29]

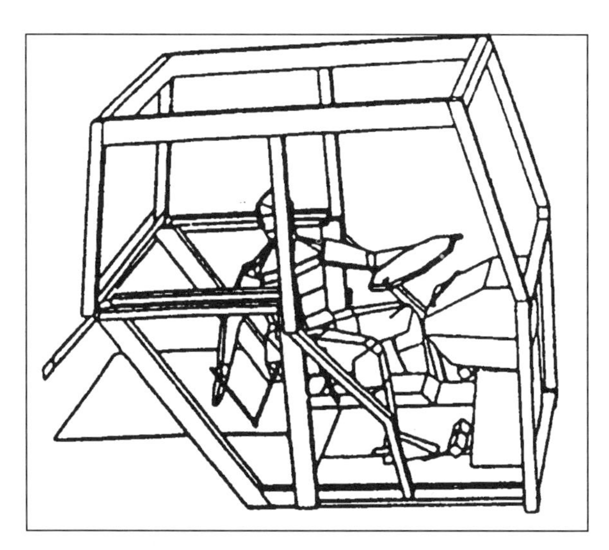

Figure 11-12. SAMMIE Link-man in a Heavy Machine Cockpit.[30]

12

Integration of Ergonomics Into Office Design

Rani Lueder

Until the late 1800s, office work assumed a very small role for most businesses. As such, office design was of little concern. Work groups and spaces were typically small, specialized needs were rare, and tasks were generally distributed evenly among workers.

Equipment such as typewriters did not become widely available until industry developed new manufacturing processes during the Civil War. Advances in mechanical servicing that accommodated deep spaces and large organizations were not introduced until the early 1900s. The rate of change was slow and easy to integrate.

Since that time, the growth of offices has greatly increased, as have their associated demands. Organizations of many sizes and specializations abound and office technologies change faster than they can be integrated into work process, office environments, and building design. Automation has introduced new and unanticipated demands on the office environment which tax existing systems beyond their limits. New materials are being used, the effects of which are poorly understood.

Members of a highly educated and sophisticated workforce are less loyal to their employer and more aligned with their professions. They expect to have a say in issues that affect them at their jobs, and they are better versed in health-related issues. Unions and regulatory activities underscore the widespread concern that new office technologies can introduce additional hazards. From another front, international competition heightens concerns that organizations remain competitive.

Such concerns have fueled an interest in learning how to make the workplace more comfortable, healthful, and productive. Also, they spur a recognition that human factors is increasingly central to office design and planning.

FORMAL AND INFORMAL PLANNING

Workspaces can be designed formally or informally. Informal planning, which is usually based on intuition, is frequently found in small companies. This type of planning often suits these companies, because they have few hierarchical levels. Management lives and works in the same environment as employees, and so understands common needs better. Consequently, communications between employees and management is more open and frequent. In these companies, the client and user organization are usually the same, and departmental objectives are similar.[1]

Larger organizations, on the other hand, may have different client and user organizations, as is common with government agencies. Davis and Szigeti[1] describe how upper management represents the client and probably occupies a different building and kind of environment than the employees, who are the users of the office design. Communication between the two is nonexistent or negligible, and many employees have little or no say about their work conditions. Facility managers have different backgrounds than the employees, occupy different spaces, and are frequently rewarded for reducing facility costs, rather than improving the work environment.[1]

Duffy, Cave, and Worthington[2] list the chief factors that can affect workspace planning in large companies, including:

- Organization size
- Large organizations usually have specialized facilities
- Rate of growth/change and number of locations
- One large building is much easier to manage than many dispersed locations
- Degree of centralization
- Managers of facilities typically have more say in centralized organizations
- Budgetary control
- Client departments pay directly to centralized areas for their space

Organizational demands in large organizations are often dynamic, complex, and in conflict. Unless these needs are formulated, decisions will likely be made be whoever takes on the responsibility or whoever carries the most weight at the time. A well-thought out program, or *user needs analysis*, helps to resolve the macro and micro issues of providing good work environments over both the short and long term. Such analyses can also help formulate directions, priorities, and alternatives for suing the business' resources effectively.

To exist within today's dynamic environment, organizational procedures typically represent a combination of formal and informal planning. In such cases, formal planning procedures may allow a larger operational structure to make informal planning decisions locally.[3]

THE ERGONOMISTS' ROLE

Ergonomists (and human factors specialists) can help organizations develop and implement new office designs as well as evaluate the office design once it is in place. Specifically, ergonomists can assist organizations:

- Ensure functional needs are met before building construction
- Formulate on-site design requirements prior to moving into a new building
- Ensure the office environment continues to support operations in anticipation of an organizational change, such as a move from closed offices to an open office plan or the implementation of new office technologies
- Resolve remaining issues once the new design is in place, which may affect employees' moral and productivity
- Conduct a post-occupancy evaluation after the transition to evaluate the new design's effectiveness, ensure user needs are accommodated, and if necessary, redefine the office design objectives
- Develop methods by which future or on-going evaluations of the new design can be compared, such as productivity improvements over time

USER NEEDS ANALYSIS

When considering a new office design or improvement of an existing one, ergonomists will often conduct a user needs analysis. The primary goal of this analysis is to identify workers' needs and how the new office design can accommodate those needs while at the same time considering the needs of the organization. One of the ways this is accomplished is to use the results of the analysis to set priorities. In turn, these priorities ensure that typically limited budgets are administered fairly and resources allocated effectively.

Ergonomists can also help a company save money by using the analysis to highlight possible consequences of the new office design. Architectural features that accommodate workers with physical disabilities, for example, only amount to 2% of a building's cost when added while the building is under construction. Later, this figure may rise to 25% of cost if retrofits become necessary.[4]

Politics within an organization are inevitable. Ergonomists can help balance all needs and interests, particularly if the workers cannot participate in the office design process as is often the case in large companies or government agencies. Additionally, if the building or office site is leased by the company, the ergonomists can also consider the owner's needs and interests.

Ergonomists sometimes find themselves in situations where there is a conflict between the perceived problem by management and what the real use needs are. For example, the working conditions of data entry operators in a large organization was experiencing a lot of physical discomfort. The reasons for health problems seemed obvious and included machine-paced workloads, high sensory demands, payment by the piece, deskilled work (typing one single line continually all day), lack of job support, and fear of job loss.

Management wanted to know what seat would cure all of their problems. Although no such seat exists, guidelines for seating selections were provided nonetheless. Additional guidelines addressing ergonomic factors of concern were also provided separately, with a discussion of the basis for and potential benefits from such changes if implemented.

IMPORTANCE OF USER PARTICIPATION

Participation is now becoming a new buzz word in industry and for good reason. Research indicates that the ability to participate at work is a primary contributor to job satisfaction and to workers' motivation to work. Being able to contribute is also related to reduced absenteeism and turnover, to improved self-esteem, and social relations with co-workers.[5]

By incorporating user input into the design and evaluation process, ergonomists not only increase the process' effectiveness, as workers often know as much as anyone about how to make their work more effective, but also increase the potential for subsequent acceptance. People inevitably resist change if imposed; yet it may even be welcomed when they help it come about.

There are roadblocks to achieving this effectively, however. Standard union procedures preclude employee involvement without going through formal channels. Timelines, as a result, may be stretched further. Those in management may resist for personal reasons, such as fear of losing control or effectiveness, and not wanting to assume greater responsibility when the normal approach is changed.

Management must also ensure that the workers' or users' input address meaningful issues. Morale may suffer, for example, if workers are only allowed to have a say in the color of carpeting. Management must also commit to accommodating the new requests if feasible.

DEVELOPING THE OFFICE DESIGN

Ergonomists apply a number of approaches to developing an effective office design.[1,3,6,7] The one discussed in this chapter is a *staged approach*, during which formal human factors criteria are incorporated into the design in stages. It also advocates tailoring these criteria to specific situations, as each organization and office design project is unique.

Getting Started

Many individuals other than the ergonomist may be involved in the office design process, including architects; designers; managers; facility managers; management consultants; vendors; union representatives; and lighting or heating, ventilating, and air conditioning specialists. Because considerable overlap between roles may exist, responsibilities must be clearly defined.

Sensitivities are often evident from the outset. The facility manager, architect, or designer, for example, may be concerned about the consequences of sharing the decision-making process or about including another specialist with an already constrained budget and timeline. Likewise, the office manager may be concerned about giving up control, such as by involving workers or union members, while retaining responsibility for office performance. The company's financial administrator may be leery that improving the office environment will not be cost-effective. Sauter, Dainoff, and Smith[8] have written an excellent book which documents convincing benefits in productivity, health, and morale from ergonomic design.

The Planning Team

To help facilitate the office design planning process, a planning team of 8 to 10 select specialists and management representatives should be appointed; keeping the team small is advisable to avoid the committee effect. The responsibilities of each team member should be specifically spelled out including each member's availability, anticipated involvement, and project responsibilities.

Although senior managers may not participate directly in the planning team, their support is essential as is the facility manager's commitment. The form of this cooperation must be clarified. Senior management, for example, must know what it wants and expects from the planning team and from the new office design itself. This information must be conveyed to the planning team as specifically as possible, along with any other necessary information relating to the project background. Finally, senior management must incorporate the planning team into the larger functional management team. A support structure must be provided to accomplish the goals established by the planning team, with someone having clear sign-off authority.

The planning team is responsible for clearly defining the office design project's goals, objectives, and scope. Elements of this definition include priorities, givens, limitation, timelines, and resources. Useful background information is needed, including anticipated change factors such as long-term organizational trends that have future design implications.

At this time senior management may, for example, decide that the project's budget will cover furniture selection but not lighting, since that is categorized under building construction and not under their office design project jurisdiction. Perhaps additional plans must be signed-off by a separate funding agency. Special interests of particular groups may be described, which make change in that arena politically infeasible.

Priorities, resources, and limitations help describe the organization's personality and how to support its inner workings. Corporate cultures differ greatly between organizations, and there will be strong resistance if such distinctions are not considered in the project definition. For some, security and privacy issues are central; for others, accessibility assumes precedence. Their image to the public, clients, or types of employees differ, as do accepted cost differentials for various environments.

Some organizations are very hierarchical and status distinctions are very important; others may be more egalitarian. When information systems are introduced, the needs of the secretary may increase from 50% to 100%, yet managerial space needs are not affected.[9]

To address such inequities, many corporations today allocate space in accordance with functional requirements, rather than by status. Further, executives are increasingly placed in the central core so that employees in the open office plan can share window views.

In some companies, work is more group and project-oriented, so a primary concern is supporting group work. Some employers may have difficulties in attracting certain kinds of workers; as a result, their design objectives differ from those of other organizations.

Workstation designs are affected by how organizational resources are allocated. When work areas lack meeting space, supplemental conference rooms become necessary. In today's dynamic business atmosphere, providing sufficient conference space is more important than ever. Meetings of all types and sizes are increasing. One facility manager recently expressed frustration that "no matter how many conference rooms we construct, we end up needing twice as many."

Printed information may be centralized or stored less space-efficiently among employees. Computers may be shared or dedicated. Technologies may be distributed or centralized in special spaces.

Other necessary information the planning team should consider includes present and projected floor plans, furniture/equipment inventories, accident/injury and health care statistics, (possibly) absenteeism rates and other data about various departmental groups, and other potentially relevant information.

Limitations are inevitable in budget, time, and resources. Consequently, decisions may be made at the workers' expense. In such cases, the ergonomist must represent end users to management by clarifying basic issues and implications of the various design alternatives. For example, management decides that moving different types of workers has become too cumbersome and expensive. As a result, they wish to eliminate individual differences in work stations by creating one or two standard *footprints* that can serve all; thus, workers, can be moved without reconfiguring the work stations. Such footprints are commonly used and necessary to make sense of the vast complexity of design considerations.[10] Oversimplifying the design process, however, may exact far greater costs than are saved. The ergonomist along with the planning team should define repercussions of such decisions, alternatives, and compromises.

The design process must also support organizational changes over time. For a number of reasons, many companies have poor documentation of future needs. Frequently, a new building is obsolete or lacks sufficient space at move-in, with no means of addressing or compensating for such inadequacies. This may occur for political reasons, due to market trend, relative growth of various departments, or limited budgets that provided little support for change.

What changes and rates of growth are expected in the organization, among departments, and groups? In three years? In five years? In the next 10 years? Will these be centrally located or dispersed? What technologies will be integrated over time? How will communication take place and with whom? What kinds of employees will be available? What kind of work will they do? Are shortages of certain workers anticipated, and how will the company attract and keep them? What steps will be taken to ensure that environmental quality standards are maintained? What about maintenance, storage, cable access? How might products, clients, legislation, and values of the community change?

This process continues in an iterative manner. From this stage a clearly defined consensus should emerge regarding the problem statement, project objectives and scope, resources, timeline, and responsibilities. A Critical Path Schedule of timelines, capabilities, and steps to complete the project should incorporate how these factors will be integrated.

Research

The planning team must then research various aspects of the project definition. This research can be conducted using a variety of methods. Generally, the more methods used, the more complete and effective the final analysis, as each method contributes information, which can help validate the planning team's project definition and implementation plan. The following are examples of some commonly used methods. The planning team should review as much information as possible, including internal company reports, published information, floor plans, equipment and furniture inventories, and other important information.

Observation may consist of informal walk-throughs by experts, formal experimental research with recorded on-site data, or techniques such as fast-action videotapes, which can provide overviews of activities or pin-point specific problems. Members of the planning team bring their own knowledge and experience to this effort. The ergonomist might note, for example, constrained postures caused by computer glare and a demanding task structure. An air quality specialist may find that plants look "ill" or that nonsmokers are down-wind of smokers, suggesting relocating workers relative to ventilation systems.

Frequently, workers provide clues of underlying environmental problems. Examples include telephone books propped under panels, suggesting pockets of dead air space caused by panels in poorly ventilated areas; home fans adjacent to clusters of equipment, suggesting excessive heat loads; telephone books placed on seats, indicating the seat does not adjust high enough or the worker is unaware that it adjusts; or taped-over windows or make-shift cardboard shields to reduce screen glare. Workers can be very creative, although their coping strategies may still exact a personal cost.

Interviews with managers and users. Interviews with managers and workers can help the planning team further define the project and special user needs. The initial interviews should be broad-based to highlight specific problems and create a context for subsequent and more in-depth investigations. Questions may address individual, group, departmental, and organizational roles; responsibilities; perceived limitations; anticipated changes; and suggestions for change. Follow-up interviews should clarify details from the general discussions in the initial interviews.

When choosing who to interview, the planning team should consider a sample of at least 1 out of 10 or 20 workers. Those chosen should be representative of the workers in their group or department. A television station's staff, for example, consists of researchers, reporters, writers, directors, producers, and assignment managers. One or more representatives from each group should then be interviewed, depending on the entire staff size as well as each department.

Questionnaires. Ergonomists recommend using questionnaires to help assess user needs. The planning team may ask interviewees to fill out a questionnaire prior to the interview. Or, if the project is not yet sufficiently defined, the interviews may introduce new questions and issues, which can be addressed in a post-interview questionnaire.

A Body Part Discomfort Scale is one type of questionnaire.[11] This scale allows workers to rate how various parts of their bodies feel throughout the day, and thereby documents patterns of discomfort over time. This questionnaire can be later analyzed along with the possible causes for the discomfort. Corlett [12] and Lueder[13] describe other, similar types of questionnaires.

Another example is a task analysis, which focuses on different types of tasks, the proportion of time spent on each, and the relationship between tasks and equipment. The results of a task analysis can show relative design priorities and placement of paper/material flow as well as suggest how to design the task structure more effectively. The following lists some sample questions, which may be included in more general user questionnaires:

- List the number of stacks of paper/documents used for any given task and their sizes. Where are the bottlenecks? These questions suggest work surface and paper flow support requirements.
- With whom do the workers communicate? How often? How important?

This information indicates relative locations and orientations.

- In order of importance, what should most be changed and why? What should definitely not be changed, and why? What is most and least-liked of their workspace, area, and/or building?
- Rate the various features of the work environment on a four-point scale from "poor/fair/good/excellent",[14] or rate the features using a range statement, such as from "strongly disagree" to "strongly agree."
- List the various features of the environment and ask how many minutes a day these features contribute or detract from worker productivity. This can help management determine the impact of the design changes.
- What are the present dimensions of workstations and equipment? The planning team can use this information to evaluate ongoing design requirements. It also indicates what furniture/equipment is presently available and what changes are necessary.
- What personal characteristics should be noted, such as bifocals, disabilities, handedness, visual problems, and allergies and asthma. What specific needs have not been considered?
- List privacy, security, and meeting requirements.

REVIEW EXISTING OR PLANNED WORK SITE

During this stage, which may occur before the research phase, specialists review specific and objective characteristics of the work environment. This may involve both measurement with specialized equipment or general comparison with criteria. This includes lighting and screen visibility, acoustics, air quality, thermal comfort, workspace dimensions, psychosocial issues and task design, and special requirements such as for workers with disabilities.

DISCUSSION

This chapter describes some of the approaches and ways in which ergonomics can be integrated into the office design process. To some extent, each corporation's culture is unique, as are their needs, interests, resources, and constraints.

Ergonomists may also be introduced at any of several project phases, from pre-design needs assessment to post-occupancy evaluation. Although management and staff support is critical, they may function alone or as part of an interdisciplinary team.

General guidelines and approaches, at best, only provide a framework to address potential problems in the organizational change. The single common denominator is that the specific needs of the users are always the central focus of the design effort.

References

1. Davis G and Szigeti F. Planning and programming offices; determining user requirements. In: Wineman, JD, ed. *Behavioral Issues in Office Design*. New York, NY: Van Nostrand Reinhold; 1986; 23-42.

2. Duffy F, Cave C, and Worthington J. *Planning Office Space*. New York, NY: Nichols Publishing Co.; 1982; 191-204.

3. Moleski WH and Lang JT. Organizational goals and human needs in office planning. In: Wineman JD, ed. *Behavioral Issues in Office Design*. New York, NY: Van Nostrand Reinhold; 1986; 3-21.

4. Ney H. Towards a barrier-free workplace. *Office Ergonomics*, 1985;2:5;23-24.

5. Lueder R. Stress in the electronic office. In: Lueder R, ed. *The Ergonomics Payoff: Designing the Electronic Office*. New York, NY: Nichols Publishing; 1986; 14-33.

6. Stokols D. New tools for evaluating facilities design and productivity. *Proceedings of the International Facilities Management Association* (IFMA); 1986; November; T-17.

7. Zeisel J. *Inquiry by Design; Tools for Environment-behavior Research*. New York, NY: Cambridge University Press; 1981.

8. Sauter SL, Dainoff MJ, and Smith MJ, eds. Promoting Health and Productivity in the Computerized Ofice; Models of successful ergonomic interventions. London: Taylor and Francis; 1990.

9. ORBIT Report. *Office Research into Buildings and Information Technology*. London: Duffy, Eley, Giffone, Worthington and EOSYS, Ltd; 1983.

10. Pulgram WL and Stonis RE. *Designing the Automated Office*. New York, NY: Whitney Library of Design; 1984.

11. Corlett EN and Bishop RP. A technique for assessing postural discomfort. *Ergonomics*. 1976;19:175-182.

12. Corlett EN. Aspects of the evaluation of industrial seating. *The Ergonomics Society's Lecture 1989*; Presented to the Annual Conference at the University of Reading, 3-7 April. London: Taylor and Francis. 1-13.

13. Lueder RK. Seat comfort: a review of the construct in the office environment. *Human Factors*. 1983;25:701-711.

Rani Lueder is President of Humanics, Inc. She has performed consulting work in the areas of seating work station and office research and design. She is a co-author of *The Ergonomics Payoff: Designing the Electronic Office* and is former Executive Editor of the magazine, *Office Ergonomics*.

13

Work Center Redesign Combines Tasks to Improve Productivity And Reduce Job Stress

Charles E. Collins
B. Mustafa Pulat, PhD

As material handling functions are being automated, many industries are implementing Just-In-Time (JIT) programs. Such efforts require disciplined, more frequent material movement in small batches, often coupled with material flow path and method changes.

These changes create opportunities to improve inefficient procedures and to apply ergonomic principles, which results in more enjoyable jobs, while at the same time eliminating or modifying potentially dangerous or stressful tasks. A redesigned handling task, for example, may improve job acceptance and reduce absenteeism.[1] The case study presented in this chapter illustrates these points and shows that redesigning a material flow path can also result in a more efficient use of existing resources.

The study is based on the board preparation operation in the kitting function at the AT&T Oklahoma City Works. The goal of redesigning the material flow path was to consolidate the four separate work centers and to reduce the extensive handling and staging required for the board preparation operation. Ergonomic principles were employed in this effort.

PARTS KITTING

A kit is a collection of components and subassemblies that together support one or more assembly operations.[2] The kitting operations at the Oklahoma City Works put together circuit board assembly (circuit pack) parts. The primary function of the kitting operation is to take separate parts from storage and picking and put them together in the form in which they will be used at subsequent assembly operations. A secondary function of kitting is final material verification.

As noted, the kitting operation occurred in four separate work centers. Figure 13-1 illustrates the material flow diagram for each of the four centers. Parts to be processed arrive primarily from the mini-load automated storage and retrieval system (AS/RS), and are verified against the accompanying material identifier information.

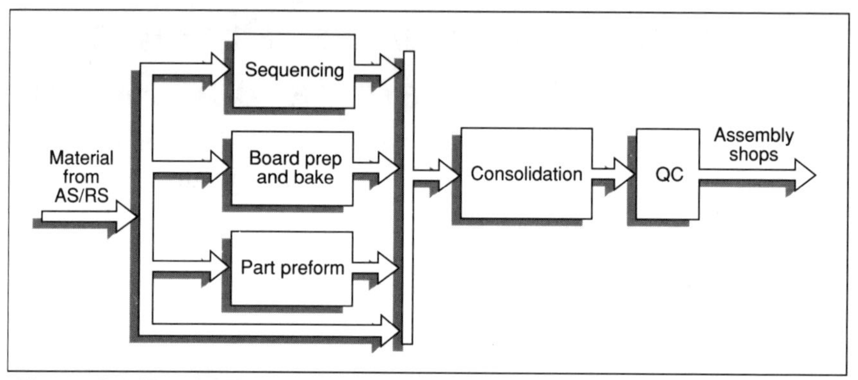

Figure 13-1. Material Flow in the Kitting Function.

Some parts will only be rearranged, so they are staged until all other subkits are complete. Other parts can take one, two, or all three routes of preparation work before being combined with the rest of the kit at the consolidation phase.

Sequencing develops an output reel of components, such as integrated circuits, resistors, and capacitors, in the order required by the insertion machines. *Part preform* involves bending, straightening, lead forming, and cutting. *Board preparation* involves stamping, cutting circuit paths, taping, drilling, labeling, and baking. The boards are baked to assure good circuit connections by removing moisture from the inner layers of board material.

The *consolidation* operation involves lot sizing, counting, stamping, and putting the assembled kits into containers. At this point, final material check is performed to make sure each kit is complete. A final *quality control* (QC) check assures kit integrity before delivery to the assembly operations.

Due to the variety of steps required, the kitting operation is notably labor intensive. Consequently, this operation greatly benefitted from the ergonomic principles that were considered when the operation was redesigned.

ERGONOMIC PROBLEMS PRIOR TO REDESIGNING THE MATERIAL FLOW

Prior to centralizing the kitting operation, four different work centers performed the same function, thus duplicating resources and efforts. As a result, each center both under and over utilized resources, which frequently generated worker complaints and at time, capacity bottlenecks.

Each center, for example, was equipped with an oven (Figure 13-2) and other board preparation facilities (Figure 13-3). This caused excessive material handling overall and discomfort, especially from the oven.[3] Also, because of the multiple sources of variability associated with four separate centers performing the same function, quality problems were extensive as was job-related stress.

Figure 13-2. Each Center was Equipped with an Oven Resulting in Discomfort.

Figure 13-3. Each Center was also Equipped with Board Preparation Facilities Resulting in Duplication and Excessive Materials Handling.

In the baking operation, workers loaded and unloaded batches of printed wiring board from the ovens. This task was physically demanding, and required workers to bend, twist, lift, and carry loads at their capability limits. Many times, workers experienced discomfort due to the heat from the ovens.

Also, workers received information related to the suboperation requirements handwritten on paper documents, called the board prep sheet and the flow tag. These handwritten documents were hard to read and were often lost, resulting in operation delays.

Overall, the board preparation operation was an excellent candidate for improvement. Redesigning the material flow not only could relieve the workers' physical discomfort, but could also significantly speed up the process by pooling resources together and organizing work into a logical sequence.

ERGONOMIC IMPROVEMENTS AFTER REDESIGN

Most of the circuit board preparation is now done at one central work center. Figure 13-4 shows the material flow diagram for the redesigned board preparation operation. As a result, 95% of the boards are processed at this one work center. The remaining 5% are processed at other work centers due to their unique work requirements.

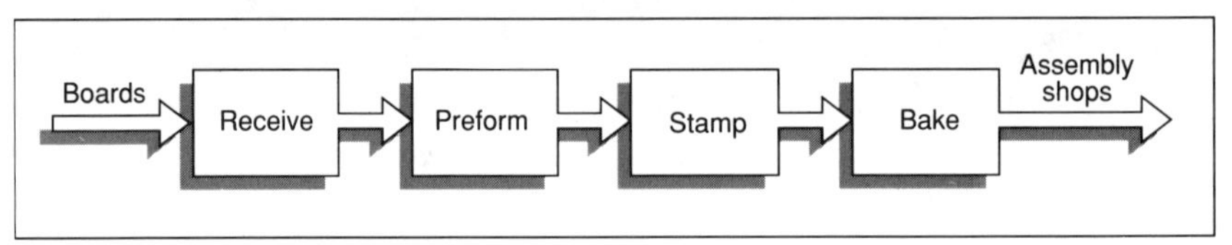

Figure 13-4. Redesigned Board Prep Operation.

At this central facility, bench operators first prepare the board (Figure 13-5). The boards are then stamped for identification and finally baked (Figure 13-6). Conveyors at knuckle height transport material from one work area to another in a serpentine configuration. Unlike the previous system, the operators intervene with the conveyor mechanisms only at two points. The entire operation is enclosed within a 52 ft by 30 ft room for additional fire control. The room has adequate space and access for material and personnel traffic.

Receiving and Bench Operations

At the receiving station, operators laser scan the boards for material tracking purposes. The boards for the same select ID, which identifies a specific manufacturing order, are routed to a particular preparation bench via a powered wheel conveyor.

Operators can request work by pushing a button, which turns on an indicator light on the instrument panel at the receiving station (Figure 13-7). The operator at the receiving station also depresses a push button, which activates

Figure 13-5. Circuit Board Preparation at a Central Facility.

the transfer conveyor and the mechanism that off-loads a board to the correct bench. Spatial compatibility between indicator light/push button positions on the instrument panel and the physical positions of the benches along the conveyor minimize work routing errors.[4] After completing the preform operation, the bench operator pushes a button to transfer the tubs containing the boards to the oven conveyor loader.

Figure 13-6. Each Circuit Board is Stamped for Identification Purposes.

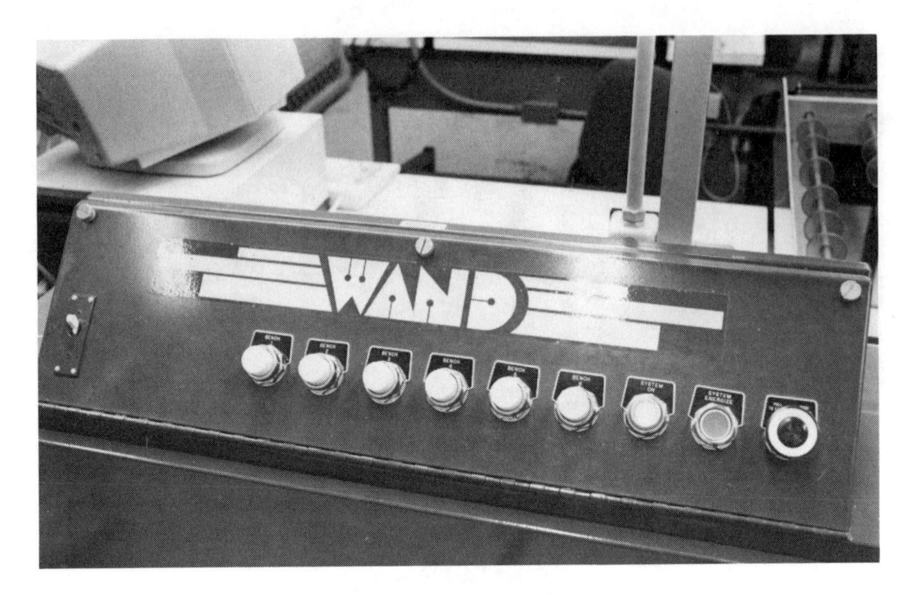

Figure 13-7. Using a System of Buttons and Lights, Operators Can Request More Boards.

Operation instructions have been automated using a CRT presentation at each position on request. Eliminating the handwritten, paper documents significantly reduced confusion. Back-up paper copies are still available for use as a contingency to the video instructions.

Automated Lift and Lower Mechanisms

The baking oven operator retrieves boards from the prep conveyor, using an automated elevator/lowerator (Figure 13-8). The operator loads the boards at waist height onto the lowerator, which automatically lowers until the stack of boards rests on an orange-tined cart placed at the base. Yellow separator boards help identify boards (green in color) that belong to different select-IDs. A manual down switch is also available to bring the lowerator down to the cart level when less than a capacity stack is loaded. When full, the operator transfers the cart to the oven conveyor loading position. For a comfortable push task, the tined cart is equipped with removable handles.

The circuit board loading device, used at the front end of the oven conveyor, is another automated mechanism used for material handling. This machine assists with loading the boards, by using a vacuum cup pick-up and push fixuring to lift the trailing edge of the boards and push them into pinch rollers onto the conveyor.

Initially, the lift device is empty and at the bottom position. If not, it can be lowered to that position via a switch. An audible alarm sounds throughout the lowering of the lift device. With the device at its lowest position, the boards are loaded between guide rails against a vertical stop. The tined cart handles are then removed and the lift mechanism started. The lift device automatically stops rising when the stack of boards is in the required position for loading onto the conveyor.

Figure 13-8. Retrieval of Baked Boards Using an Automated Elevator/Lowerator.

Stamping and Baking

Prior to the baking operation, the boards are date coded by a date code stamping machine mounted to a conveyor section attached to the loader. The boards are then fed into the flowthrough oven by the conveyor and right-angle transfer mechanisms.

After baking, the boards are stacked on an automated drop leaf stacker. The stacker automatically activates a light and sounds an alarm when the stacker is full and needs operator attention. At this point, the operator removes the boards, groups them by select-IDs, and sends them to the assembly operation.

CONCLUSIONS

The redesigned work flow at the kitting work centers has eliminated all material handling except that required to load and unload the conveyor at specific points. Now, all handling is done in a standing position, with no stooping, bending, or reaching. Prior to this redesign, workers had to lift tubs of boards to and from portable cars, work benches, and shelf racks. They also had to travel excessively to stage the product in kits.

Using automated equipment operated by programmable controllers, all linked together for in-line continuous operation, while retaining operator intervention features has increased productivity beyond expectations. The control consoles on each bench for signalling the need for more work and for transferring finishing work onto the conveyor allows the operators to be a part of the automation project and feel in control of it.

Work is still in progress for further automation of information between work areas. CRT-video presentation is being enhanced both functionally and ergonomically.

Overall, this redesigned system allows for a more enjoyable, acceptable, and comfortable work environment. At the same time, resources are better used by consolidating capacity. The material is moved faster through the process, with a resulting drop in process inventory.

References

1. Pulat BM. Introduction. In: Alexander DS, Pulat BM, eds. Industrial *Ergonomics: A Practitioner's Guide.* Norcross, GA: Institute of Industrial Engineers; 1985;1-7.

2. Bozer, YA, McGinnis LF. *Kitting: A Generic Descriptive Model.* Atlanta, GA: Georgia Institute of Technology; May 1984; Technical Report MHRS-TR-84-04.

3. Ayoub MM, Selan JL, and Jiang BC. Manual materials handling. In: Salvendy G, ed. *Handbook of Human Factors.* New York, NY: John Wiley and Sons; 1987;790-818.

4. McCormick EJ and Sanders MS. *Human Factors in Engineering and Design.* 5th ed. New York, NY: McGraw Hill Book Co.

Charles E. Collins is a retired department head from AT&T where his major accomplishments were the development of single slot coin telephone, exhaust free freon cleaning process, and various automated material handling installations. Mr. Collins is also served in and retired from the marines where he achieved the rank of full colonel. Mr. Collins is a registered professional engineer and received his BS and MS degrees in mechanical engineering from the University of Oklahoma.

14

Ergonomics of a Packing Work Station

Stephan Konz, PhD
Larry Noble, PhD

In April 1986, the ergonomics subcommittee at a food manufacturing plant contacted the investigators concerning the musculotendinous problems of the packers in the packaging department. After visiting the work site and meeting with the plant management, both parties agreed to conduct a multi-faceted analysis of the packaging operation, with the goal of developing recommendations to minimize the musculoskeletal and tendinous injuries.

During the following month, the investigators visited the plant on two occasions to observe, collect data, and perform the analysis. During these visits, the size of the packers, the packing station dimensions, and layout were measured. Also, the job and packing station were video taped to allow for a more detailed analysis. A third visit was made to interview the videotaped packers and to make additional observations. From the results of the analysis, recommendations were developed in the following areas:

- Modifying the packing station configuration
- Implementing a program to train all packers in the most efficient movement patterns
- Applying relevant factors in selecting packers
- Applying training, pacing, and rotation to prevent undue stress and injury
- Keeping records on variables relevant to injuries

EXISTING SITUATION

The packers on each of three eight-hour shifts normally work five days each week. Each packer actually works 420 minutes per shift, with two breaks of 15 minutes each (paid) and a lunch period of 30 minutes (unpaid). All packers may have to work up to four hours overtime each day when production demand is high. Each packer packs about 6,000 boxes per week, each with 10 to 16 bags. Packers shift every hour to a slightly different bag on a different line.

All packing stations have essentially the same configuration and design, with slight modifications to accommodate bags and cases of varying sizes. Figure 14-1 shows the generic packing station, which includes an incoming conveyor, a box stand, a carton dispenser, a label dispenser, and a takeaway conveyor. The job is to take bags of various sizes from an incoming conveyor and place them in cardboard boxes to be shipped to retailers.

The packers are responsible for:

- Inspecting the finished bags for correct price, date, seal integrity and appearance
- Placing the correct number of bags in the correct size box with the proper label attached to the box
- Maintaining an acceptable sanitation level throughout the shift and assist with shift change sanitation tasks

In all product lines, the packers used the following job methods to complete the packaging procedure:

(1) Remove carton from the carton drop
(2) Fan fold bottom of carton
(3) Attach label to correct location on the box
(4) Place bags in box in the desired arrangement,
(5) Fold top of box
(6) Aside finished box on take-away conveyor
(7) Drop cartons and perform housekeeping tasks as necessary throughout the preceding cycle

There are numerous variations in the technique of completing each method and in the timing of label placement. The job requirements are considered to be met if a packer maintains pace with the bag machine, maintains an acceptable sanitation level in the area, and helps ensure that only bags of acceptable quality are packed.

Nature and Incidence of Injuries

The primary problem was the incidence of overuse injuries to the upper extremity, including finger and wrist flexor tenosynovitis (FTS), carpal tunnel syndrome (CTS), and shoulder-joint overuse syndrome. During 1985 and 1986,

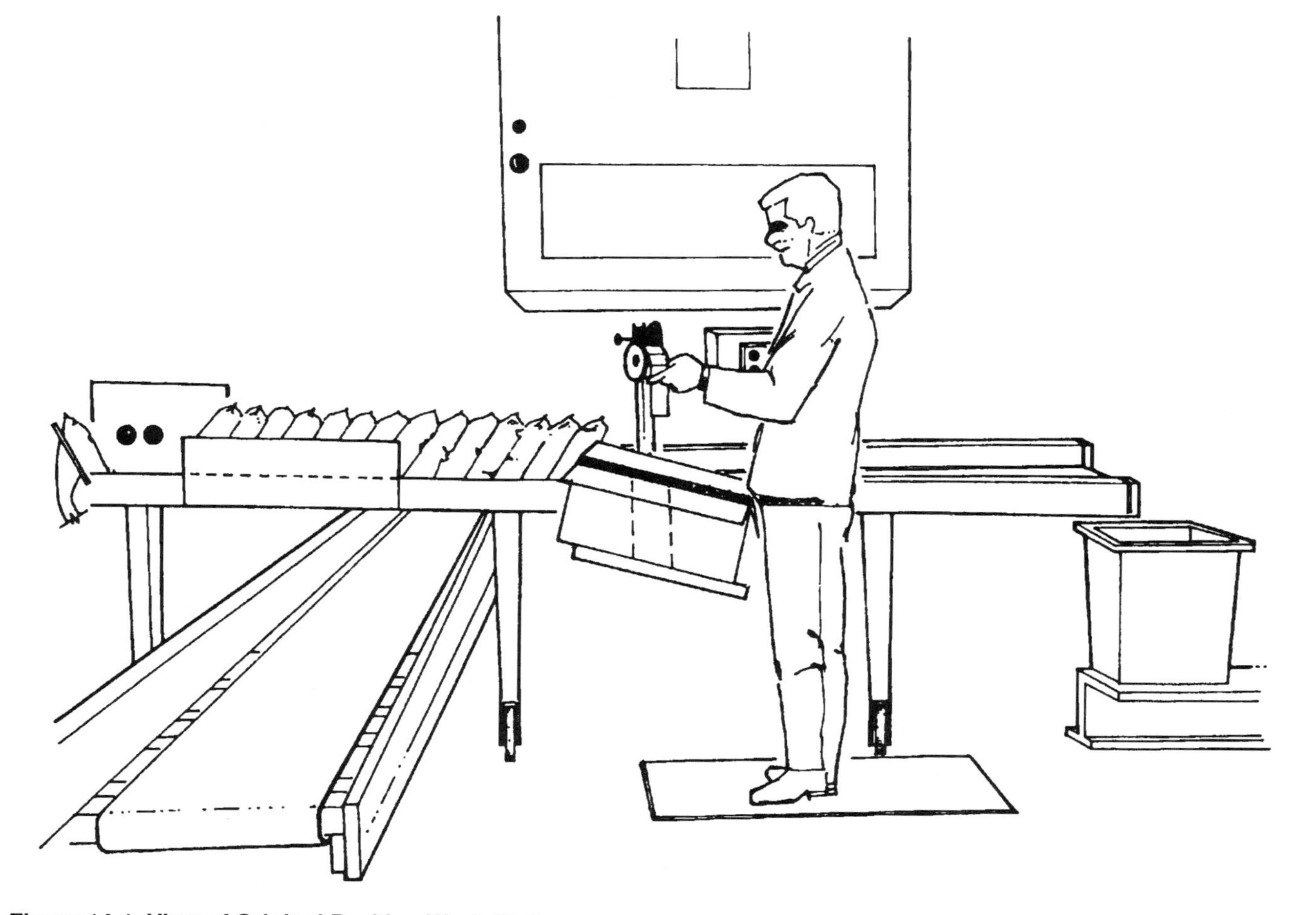

Figure 14-1. View of Original Packing Work Station.

a total of 59 and 45 injuries, respectively, were reported in the packaging department (Table 14-1). In 1985, 10 of the 15 hand and wrist injuries were diagnosed as FTS of hand and finger flexors and/or CTS. All of the injured packers were women, ranging in age from 25 to 46 years (mean + 32.9 years). Tenure on the packaging line of those with CTS ranged from 2 to 10 years (mean = 5.0 years). In 1986, 9 of the 14 hand and wrist injuries were diagnosed as FTS of the hand and finger flexors and/or CTS. Also, a substantial increase in shoulder joint injuries was noted in 1986.

Table 14-1. Injuries in the Packaging Department During 1986.

	Number of Injuries		Percent	
Body Part Affected	1985	1986	1985	1986
Shoulder	2	7	4	16
Arm and forearm	9	4	16	9
Hand and wrist	15	14	26	31
Back	9	7	15	16
Other	23	13	39	29
Total	59	45	100	100

This project began in June 1987. Since January 1, injury reports already showed 16 suspected CTS injuries in the packaging department. Ten of these injuries occurred in women, and 9 of the 16 injured packers had been working in packaging two years or less.

Selection and Training

Management routinely placed most new production employees in the packaging section. There were no specific criteria for the job. Approximately two-thirds of the packers are women. Most employees remained in the packaging section for less than three years. Many of them quit for various reasons, and some transferred to more attractive positions in the plant.

The existing packer training program consisted of one work week of the following activities:

- Day 1: Observe packer trainer and others for 30 minutes and read packer training manual. The manual included many topics such as: plant layout, production process, production procedure, product inspection, paper work, safety, sanitation, and injury prevention techniques.

- Days 2 through 5: Work the package line at regular production speed for four hours each day. Continue to study training manual.

During the entire week, the packaging trainer was available periodically to answer questions and offer advice. Co-workers were encouraged to assist new packers when they got behind on the production line.

Packers rotated from one line to another each hour on the hour. While training literature encouraged rotating from large bags to light bags and from left side conveyor to right side, this policy was not followed.

Preconditioning and Fitness Exercises

No systematic guidelines or programs involving preconditioning or work hardening exercise regimens were available to packers, regardless of experience. Some training materials included suggestions for flexibility and tension reduction exercises to be completed during the workday. These materials, however, were not made available to all packers in a systematic way.

Notable Injury Prevention Procedures

The ergonomics subcommittee encouraged packers to use elastic wrist bands to help prevent CTS and FTS. Several packers were observed using them. The committee was considering suggesting floor mats at the time of the first plant visit at one or two packaging lines and a policy to guide packers on appropriate footwear.

Also, the committee encouraged packaging techniques using both hands to pick up bags and the *scoop* method in hopes of preventing CTS. Previous experience suggested that repeated *pinching* to pick up bags with one hand causes CTS. Techniques involving reaching and rotating and flexing the spine at the same time were discouraged.

ANALYSIS METHODS

The analysis of the packing station and the job consisted of the following steps:

1. Systematic observation of each packaging line from several viewpoints. During this procedure, results of observations and ideas were written down and later reviewed.
2. Video tapes of packing procedure. We chose to video tape one of the easier product packaging lines. Five packers representing different gender, height, and experience levels, were video taped. Each packer was taped for several packaging cycles from three different camera positions, top front, side, and rear, and using varying focal lengths, such as wide angle and close-up. A 20-minute edited video tape was produced from taped footage for detailed analysis. Each packer was interviewed and asked the following questions:

- Have you had any physical ailments resulting from work, which prevented you from working?
- If yes, what body part(s) was (were) affected?
- Do you have any suggestions for improving your training program?
- Do you have any suggestions for changing the packing station?
- Do you warm-up prior to beginning work?

3. Measurements of the packer and packing station. The following measurements were taken:

- Packer stature, elbow heights, and shoulder height
- Height of packing station elements
- Horizontal distances between packer and various packing station elements

Relevant ergonomic and biomechanical principles were then applied to analyze these data. Recommendations did not involve any change in job responsibilities or production rate, but focused on:

- Redesigning the work station
- Eliminating unnecessary movements
- Minimizing segmental acceleration, such as jerky movements and abrupt change of speeds and direction
- Minimizing segmental acceleration, such as jerky movements and abrupt change of speeds and direction
- Minimizing the physical work necessary to complete the packaging task
- Eliminating the following movements:
  - Pinch grip
  - Extensive wrist flexion
  - Wrist hypertension
  - Raising elbow to and above shoulder level
  - Simultaneous spinal flexion and rotation
  - Extensive lumbar spinal flexion
  - Lumbar spinal hypertension

Implementing a systematic packer training program. The goal of this program is to assist both novice and experienced packers in adopting preferred packaging techniques (Table 14-2).

SOLUTIONS

Carton Drop

Originally, the bottom of the carton drop was 147 cm above the floor. In addition, the packer had to lift the carton above a 10.2 cm front lip, making it

Table 14-2. Suggested Packer Training Schedule.

Day 1	Day 2
2.0 Hours orientation into the plant 2.0 Hours packing with trainer	1.0 Hours with packing manager 1.0 Hours with tape and description 2.0 Hours packing with trainer
4.0 Hours total	4.0 Hours total
Day 3	**Day 4**
4.0 Hours packing with trainer	4.0 Hours packing with trainer
4.0 Hours total	4.0 Hours total
Day 5	**Day 6**
7.5 Hours total Packing with trainer*	7.5 Hours total Packing with trainer (optional)*
Day 7 Start regular schedule	
7.5 Hours total No trainer*	

*Tape review and training station as necessary.

necessary to clear a minimum height of 157 cm. Thus, most packers had to reach above their shoulders to get the carton. The drop could not be relocated and had to stay above the conveyor.

The bottom of the carton drop was lowered to 122 cm above the floor. A lower position, with the drop still above the conveyor, would not permit bags to pass under the drop. In addition, the bottom of the carton drop was modified so the cartons were presented in a standardized position.

The drop floor was slanted toward the packer so the cartons were in the side of the drop closest to the packers. The ends of the drop floor also were slanted so the carton was centered in the drop. Additionally, the worker placing cartons in the drop (on the floor above) was instructed to use a consistent carton orientation, which simplified the orientation task of the packer when removing cartons.

Supply Conveyor

Ideally, the packer should work with an object about 5 cm below elbow height. Elbow heights naturally varied considerably among the packers. In addition, the plant had a policy of rotating packers every hour to vary muscle groups and to distribute easy and tough tasks. Thus, management did not consider it feasible to adjust the conveyor heights to specific heights for each packer; however, placing the conveyor to suit the average packer was feasible.

When packer elbow heights were measured, the conveyor was 13 cm to 30 cm below the elbow. Since packers were grasping the bags about two inches above the conveyor, the grasp was 8 cm to 25 cm below the elbow, which required excessive bending. Consequently, the conveyor was raised from 84 cm to 96 cm. Because the conveyor had the adjustment built into the supports, no capital cost was required for the change.

Box Stand

The formed cardboard box was placed on a box stand, which varied in height from 53 cm to 63 cm. As an interim measure, the stand was set at a height of 61 cm. Then corporate engineering designed a new box stand with the following features:

- Easily adjustable height by the packer, that is within 5 cm
- Easily adjustable slanting top
- Top designed to reduce box movement (short lips on sides and high-friction surface)

This design required the engineering staff approximately one-year to develop. At first they tended to make the design too elegant and not rugged enough.

Box Disposal and Overflow

Also during the original procedure, if the bags were not packed by the packer they continued to the end of the conveyor and fell into overflow bins. When time permitted, the packers reached into the bin and removed the bags and packed them into cardboard boxes. The bottoms of the bins were only 15 cm above the floor, thus requiring an awkward, stressful reach. The height of the rack holding the bins was modified so the bottoms of the bins were 38 cm above the floor.

Label Dispenser

The label dispenser was moved so the reach for the label was about 15.1 cm less than the original placement. An attempt was made to use a dispenser that would automatically peel the label from its backing paper and eliminate the pinching action needed to remove the label from the paper. The machines furnished by vendors, however, did not prove sturdy enough for the production operations.

Floor Mats

Prior to the project, the packers stood on concrete. As the plant produced a food product, management was very concerned with sanitation and vetoed carpet. After some searching by the ergonomics subcommittee, a 91 cm x 122 cm rubber mat was purchased for each packing station, which can be moved and cleaned each day.

Table 14-3. Closing the Flaps on a Cardboard Box (see Figure 14-1).

Left hand	Right hand
Bad: Tucking the corner away from body stresses shoulder.	
Left hand	Right hand
Reach to 1B	Reach to 2R
Grasp 1B	Grasp 2R
	Fold 2 flat under 1
Fold 1 flat over 2	
Release 1	Release 2
Reach for 4L	Reach to 3
Grasp 4L	Grasp 3
Fold 4 flat over 1	
	Fold 3 flat over 2 and over 4
Release 4	Hold 3 flat
Reach for 2R	
Grasp 2R	
Pull 2R above 3F	
Release 2R	Release 3
Good: Tucking in corner close to the body allows better leverage.	
Left hand	Right hand
Reach to 2	Reach to 3B
Grasp 2	Grasp 3B
	Fold 3 flat under 2
Fold 2 flat over 3	Release 3
Release 2	
Reach to 1	Reach to 4
Grasp 1	Grasp 4
Fold 1 flat over 2	Fold 4 flat over 3 and over 1
	Release 4
Release 1	Reach to 3B
Reach to 4	Grasp 3B
Hold 4 flat	Pull 3B above 4R
Release 4	Release 3

Note: This activity should be done close to the body to reduce repetitive strain on the shoulders. Use four playing cards or rectangular pieces of paper to visualize the pattern.

PACKAGING METHODS

The video tapes of the five different packers showed that the packers used a great variety of techniques for making the box, grasping the bags to put into the box, and packaging patterns within the box. The authors analyzed the methods and evaluated those that caused more or less strain to the packer. Table 14-3 and Figure 14-2 show the recommended method of fan folding the carton flaps. The technique to avoid is inserting the last flap on the side of the box away from the packer as this requires a stressful exertion at the end of a long moment arm.

Grasping the bags is best done with a scoop grip in which the thumb and fingers do not pinch toward each other. A stressful technique is to grasp two or more bags by pinching them between the thumb and first finger. The problem is not so much excessive force as the 25,000 or more repetitions/week, assuming three to four bags per pinch.

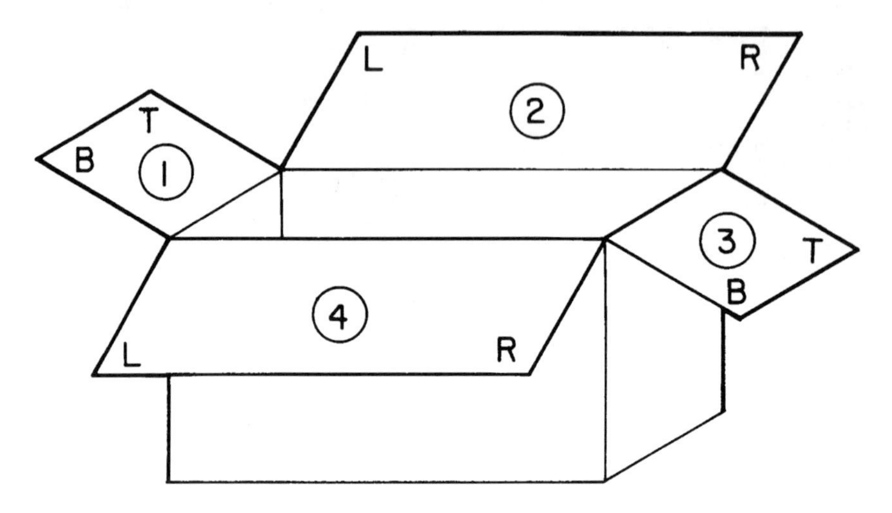

Figure 14-2. Isometric View of Box with Flaps Labeling System.

A training video tape was made showing the proper method of making the box, fan folding the box, packing the box, and disposing of the box as well as grasping the bags. One of the points emphasized in the training tape is to promote variety in micro-methods so that different individual muscle groups are used and repetitions per muscle group are reduced.

A training program was also recommended, which advocated breaking the packer in gradually, with initial days spent working at the low difficulty stations. Packers would be given specific instruction and practice on each part of the packaging procedure without the pressure of production speed. Reviewing video tapes of trained and experienced packers also were features of the recommended training program. Only after some experience would the new packer be moved to the high difficulty stations. Overtime would be restricted for novices until they had become both more skilled and more work-hardened. A number of simple stretching and flexibility exercises were given to the packers as well.

RECORDKEEPING

The authors recommended that records of the following items be kept for all injuries resulting in *restricted* jobs:

- Date and location of injury
- Identity of injured worker, including name, age, gender, height, weight, experience on job
- Nature and extent of injury
    - Chronic or acute
    - Body part affected
    - Diagnosis
    - Days of work
    - If no work missed, extent of job restrictions
    - Suspected work task causing injury
- Exercise habits of injured worker

The data would be useful in further characterizing the risk factors involved and evaluating the effectiveness of the prescribed methods.

Unfortunately, many work analyses focus solely on ways of reducing time per unit. While this focus should not be ignored, it is a short-range strategy. Reducing stress and making the job easier permits more people to achieve standard time and with less practice. In addition, less stress can reduce pains and strains, which in addition to being desirable on ethical grounds, can have large financial rewards in the reduction of injuries, such as carpal tunnel syndrome, shoulder problems, and back problems.

Technology has provided two new tools for job analysis: the video camera and the video tape recorder. Box 1 presents tips for video taping analysis. The low capital cost and the ease of using these tools allows every facility to use them.

Changes can be more easily implemented if other parts of the organization support them. Stress reduction and improved safety and health are popular with everyone. In this project, the authors had strong support from the ergonomics subcommittee of the safety committee and from the vice president of manufacturing who expressed a public commitment to improve working conditions. This support helped reduce barriers to change and significantly improve the effectiveness of the changes.

A formal program to review jobs periodically may bring into light obvious changes required such as modifying distances and reducing spinal twisting and turning. Then, using the Pareto concept, most attention can be given to the improvement of the "significant few" rather than the "insignificant many."

Box 1. Video Taping Jobs

Video taping jobs is useful for methods analysis and training, and has replaced the use of movie film due to the low capital and tape cost, its ability to be used with available light, and its instant replay capability. Formerly black and white TV cameras, were used due to the high cost of color cameras, but color now is the standard. Although most cameras take satisfactory pictures in normal factory lighting, a low-light capability is handy as is the ability to shoot a high number of frames/second, such as slow-down motions.

Some Shooting Tips

- Study a variety of operators if possible. Small variations in technique become noticeable on tape that are hard to detect by direct observation. For example, what sequence of steps is being used? Is the wrist twisted or straight? How is an item oriented?
- Plan the location of camera and subject ahead of time. Multiple views are best. Consider some combination of a front view, a side view, a back view, floor level views, stepladder or partial plan view, overall views, and closeup views.
- Use long shots. Each scene should have several cycles if possible. For methods analysis a lot of repetitions is preferable. Later, if desired, for training or management presentations, an edited version can be presented. Use a tripod. Begin the scene with a full view (far shot, wide-angle view) and then zoom in as desired.
- Audio can be used as a "notepad" while filming. For presentations or training, dubbing in a voice reading a script may be desirable.
- After shooting, have both the ergonomist and the worker analyze the tape. The video cassette recorder should have freeze-frame, single frame advance, and a remote control. The worker can point out to the ergonomist why some things are done, perhaps with audio dubbing. The tape also is a way to show the worker some problems.

Stephan Konz has been a full professor in the Department of Industrial Engineering at Kansas State University since 1969. Dr. Konz has authored five books and has 160 other publications to his credit. He earned his PhD in Industrial Engineering from the University of Illinois.

Larry Noble is Professor and Head of the Departent of Physical Education and Leisure Studies at Kansas State University. He received his PhD in Physical Education from the University of Texas at Austin, specializing in Biomechanics and Exercise Physiology. He is also Director of the Biomechanics Laboratory at Kansas State University and has been involved in research in the general areas of physical fitness and athletic performance. His analysis of these activities has involved instrumentation, cinematography, and other imaging techniques as analytic aids.

15

Investigation of the Automobile Spare Tire Load Operation

Steven Johnson

Workers who assemble automobiles perform many repetitive movements during their shifts, some of which contribute to cumulative trauma disorders (CTDs) of the wrists and can cause lower back problems. An example of an assembly operation that falls into this category is spare tire loading. The investigators targeted this operation for three reasons:

- A history of reported medical visits and worker's compensation claims
- Worker complaints about fatigue
- Possible use of the jack installation procedure in another product

Workers alternate at one-hour intervals between two different configurations of the spare tire load operation, one of which involves a side-loading of a smaller tire along the inside of the right rear fender. The other configuration involves mechanically delivering the normal-sized tire to the top of the right-rear fender. The procedure used during the second configuration follows:

1. Roll the tire into the trunk
2. Lift and place the tire to the front center of the trunk
3. Assemble the jack plate, J-hook, and wing nut
4. Attach the J-hook through the tire hub to the trunk
5. With an air wrench, tighten down the tire (wing nut)

According to the job description, the workload factor for installing the smaller tire is 97.5%, whereas the workload factor for installing the normal tire is 70.2%. Even though installing the normal tire required less of a workload in terms of time and allowed the workers more recovery time, this configuration caused greater worker fatigue and back strain.

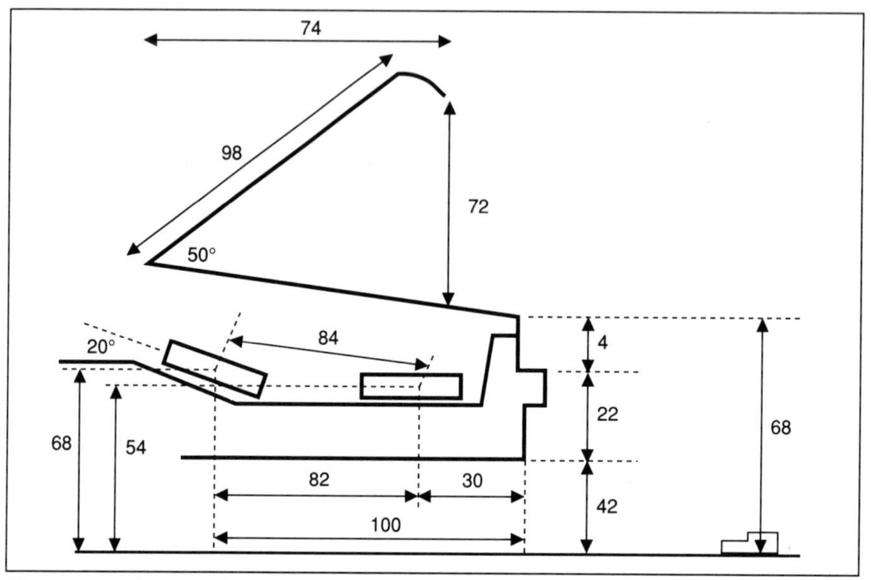

Figure 15-1. General Geometry of the Arrangement.

BIOMECHANICAL ANALYSIS

Initially, the investigators thought that bringing the normal tire from the rear to the front of the trunk caused the fatigue and back problems. Figure 15-1 illustrates the general geometry of the trunk-tire arrangement. If this was the case, then some type of mechanical assist might prevent the problem.

First, the investigators evaluated the job relative to recommendations provided by Stover Snook[1] that have been incorporated into the National Institute of Occupational Safety and Health (NIOSH) document titled, *A Work Practices Guide for Manual Lifting.*[2] When considering the height, distance moved, frequency, and weight of the tire, loading the tire should not have been a problem for the work group assigned to the job. That is, approximately 90% of the male workers should not be a risk for injury when performing this job.

Second, the investigators used a biomechanical model developed at the University of Michigan to determine the compressive forces on the lumbrosacral area (L5/S1).[3] Figure 15-2 illustrates the geometry of lifting the tire relative to the compressive forces on the L5/S1 area. Using the geometry of the tire moving task and the characteristics of the worker, the compressive forces were calculated. Again, as with the inspection of the tabulated values, the compressive forces were not strong enough to cause a high risk of injury.

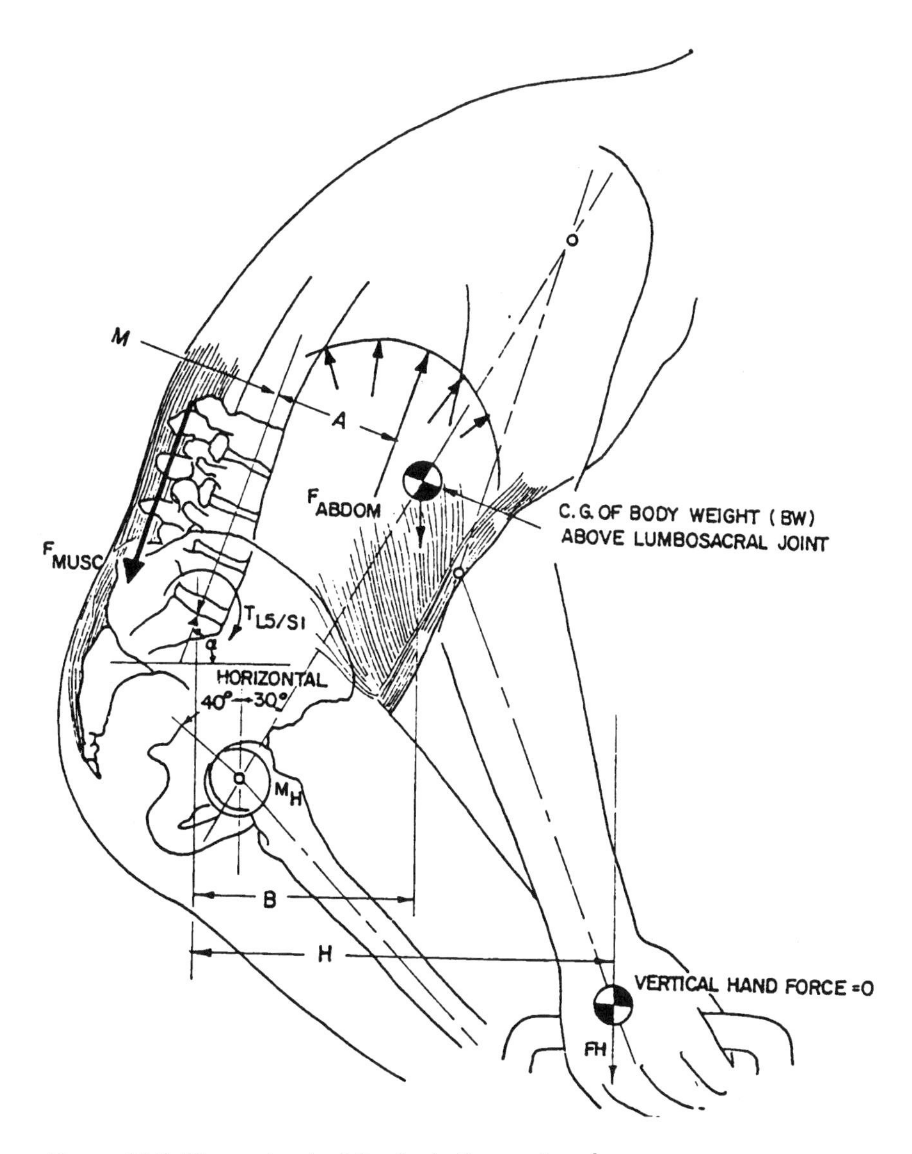

Figure 15-2. Biomechanical Analysis Parameters.[3]

From the results of these analyses, the investigators concluded that the actual manipulating of the tire was not the primary cause of the complaints and injuries. Therefore, the other portion of the workers' job was addressed, specifically attaching the tire to the truck floor. This task required workers to assemble the J-hook, jack plate, and wing-nut, with their torsos nearly horizontal, resulting in significant strain on their lower backs.

When the investigators compared the two portions of the job, loading the tire and tire tie-down, the compressive forces were actually slightly higher for the tie-down portion of the job. These calculations were based on a static model of lifting in the frontal (midsagittal) plane. The geometry of the tire tie-down task is consistent with this analysis, the dynamic forces were not addressed, however.

The movement of the tire, and the resulting forces occur for a relatively short period of time. The tire tie-down task, which is performed in a poor posture, takes much longer. Even though the compressive forces are similar, the work required in the two situations is quite different, with the tie-down task requiring much more work than the tire manipulation. The biomechanical analysis supports this view. The actual loading of the tire was not the problem; but rather, the workers' assembling, tying-down the tire, and lifting their torsos out of the trunk caused more serious lower back problems.

ELECTROMYOGRAPHIC ANALYSIS

When a muscle contracts, a large number of individual nerve fibers, called *motor cells*, act together as a *motor unit* to stimulate the muscle fibers. In response to the contraction, the muscle discharges a detectable electrical charge, called an *action potential*. This discharge, expressed in millivolt units, can be measured with surface electrodes placed over the muscle.

Because the more force that a muscle exerts, the greater the recorded electrical activity. One way to investigate this is to evaluate the exertion and consequently, the fatigue of a muscle or muscle group, by measuring the muscles' electrical activity during work. If tying down the tire is more stressful than loading the tire, then the cumulative electrical activity generated by the workers' muscles during that portion of the job should be significantly greater.

To investigate the relative effect of the tire load versus the tire tie-down tasks, two different tire manipulation methods were studied. Some workers grasped the wheel hub with the right hand, lifted the tire with that hand, and moved and oriented the tire with the left hand. Other workers grasped the outside of the tire on opposite sides and lifted, moved, and oriented the tire with both hands.

In the latter case, the load during the lift is symmetrically distributed on both the left and right side of the back, and is characteristically smoother and involves less jerk during the lift. As expected, the evenly distributed method resulted in less muscle stress and less potential for injury.

The investigators also evaluated the relative muscle stress caused by a proposed method for installing the jack. In particular, the stress involved with the jack installation was compared with the spare tire load operation. If workers experienced greater muscular stress when installing the jack than when installing the spare tire, then the recommendation to develop another method for installing the jack would be supported. The simulation of the jack load operation was not entirely accurate, because at the time the exact location of the jack was unknown, and the fixture used to hold the jack did not yet exist. The results of the simulation indicated potential problems, but a more valid study was needed to adequately evaluate the design.

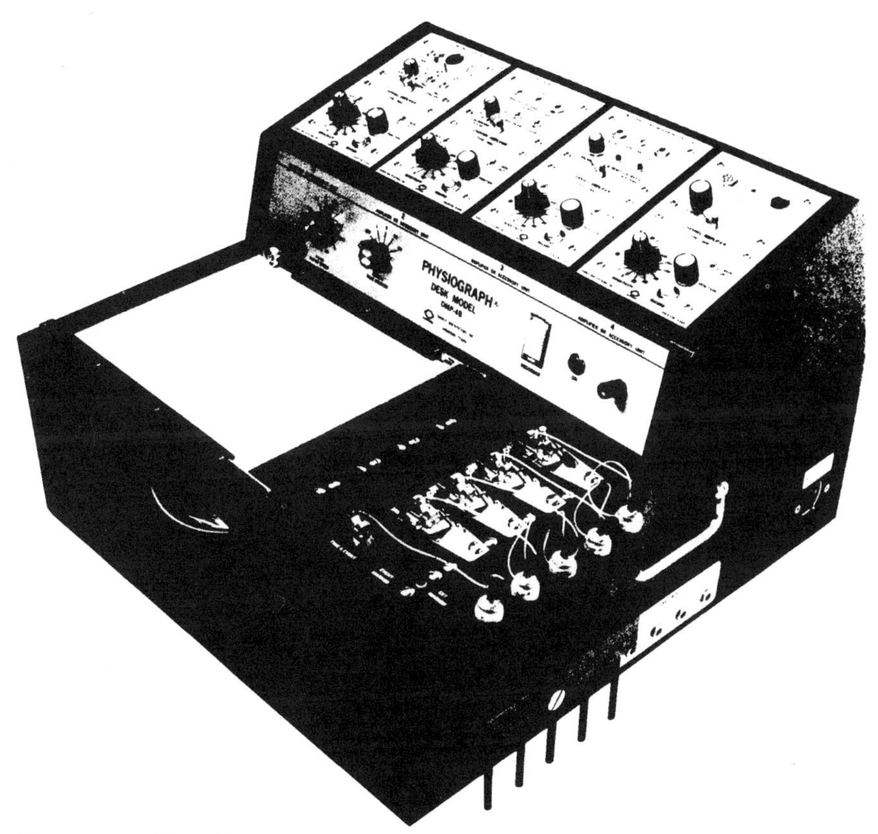

Figure 15-3. Four Channel Physiograph.

ELECTROMYOGRAPHIC EQUIPMENT

The investigators primarily used a portable four-channel physiograph (Figure 15-3). This instrument amplified, filtered, and displayed the electromyographic (EMG) signal from the workers' lower back muscles. Three surface electrodes (reference, active, and ground) were placed on the workers' lower back. The active electrode was placed over the muscle adjacent to the fourth lumbar vertebrae, the reference electrode was placed 2 in away at the second lumbar vertebrae (45° angle to reduce interference from action potentials generated by the heart), and the ground was placed midway up the back and close to the spine.

The amplified signal was rectified and integrated. The integrated EMG signal depicted the electrical muscle potential as a cumulative measure of each portion of the job; tire manipulation versus tie-down. The integrated muscle potential measure was plotted. The amplified signal also went from the physiograph to a *signal smoother*, which detected the peak values of the electrical response. This measure was also plotted.

The raw amplified signals, as well as the integrated and smoothed signals, were also recorded on a four-channel (plus voice) analog data recorder. The recorded data were then returned to the laboratory for a more detailed, off-line analysis.

The investigators spent three days setting up and calibrating the equipment and scaling the signals. During the initial portion of this effort, they attempted to use a telemetry system so that the workers would not be hindered by wires connecting them to the instruments. The ambient electrical noise prohibited the use of the telemetry system, however. While gathering the data, the wires between the workers and the instruments did not pose a problem. Portable systems, such as recorders and play-back units can be used when a direct connection is undesirable.

RESULTS AND CONCLUSIONS

Spare Tire Load Analysis

Figure 15-4 illustrates a typical recording of the spare tire load operation. The lower trace is the raw, amplified muscle potential. The upper trace is the integrated, positive only signal. The raw signal shows that, although the initial task (tire manipulation) is more extreme, the tie-down task (particularly the first part) also results in large potentials. All 26 trials showed that the integrated signal, the cumulative effect (effort over time) is more severe for the tie-down task than for the tire manipulation. The average cumulative potential for the tie-down task was 220% of the tire manipulation. That difference is both statistically and practically significant. The slopes of the traces indicate that the time involved in each component of the task is the prime contributor to the cumulative effect.

This analysis supports the initial evaluation based on accepted standards and biomechanical analysis. The assembly of the tie-down bolt, jack plate, and wing-nut, along with tightening with the power tool and the workers' lifting their torsos out of the trunk, resulted in more severe muscle stress than the actual manipulation of the tire.

As expected, the use of an evenly distributed load on both sides of the body resulted in lower cumulative muscle potential. Interestingly, this was not the expected result based upon the workers' subjective evaluation. One possible explanation is that, although less stress is exerted on the lower back in the distributed case, the hands, wrists, upper arms, and shoulders play a much larger role in manipulating the tire. The overall exertion, therefore, may be comparable, even though the stress on the lower back is reduced.

Jack Load Analysis

Figure 15-5 illustrates the raw and integrated signals recorded during the jack load task. A comparison of these data with the signals recorded during the spare

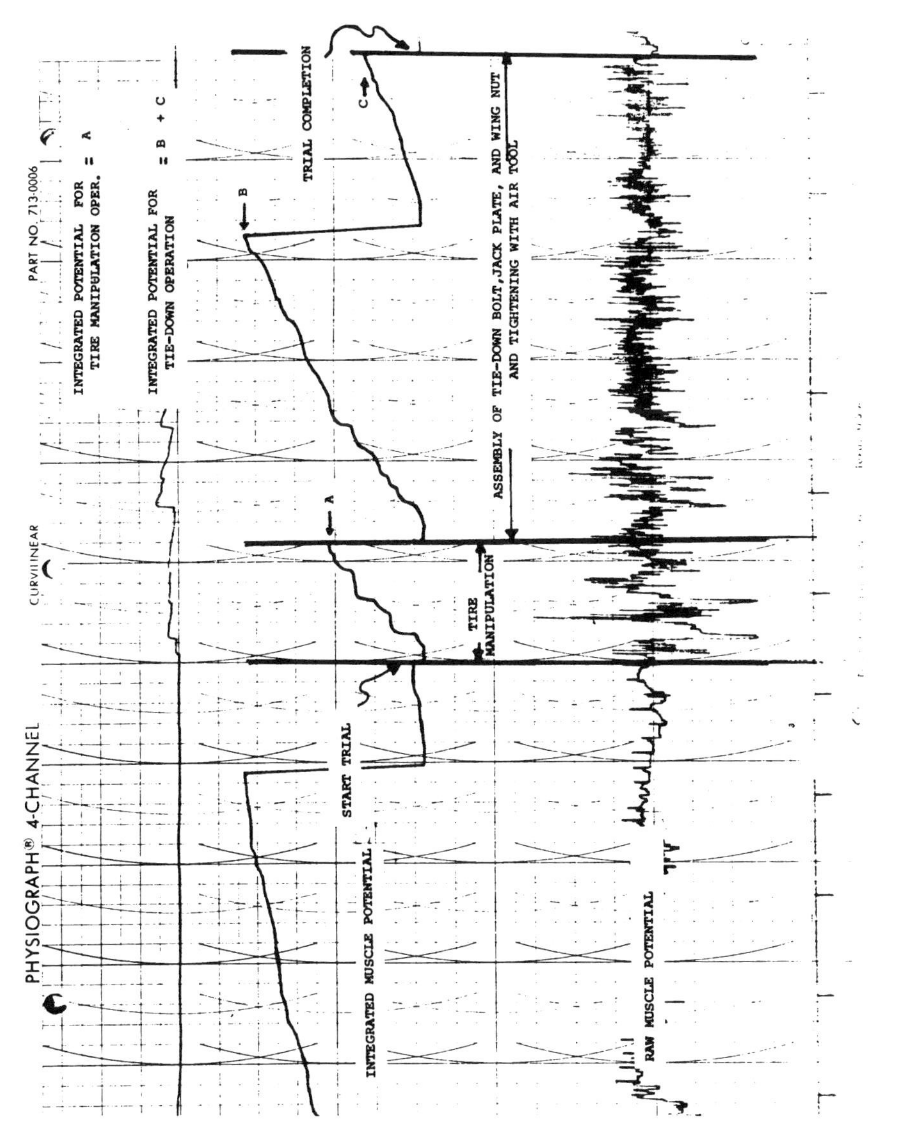

Figure 15-4. Electromyographic Recordings for the Spare Tire Load Operation.

tire load operation shows that the loading of the jack behind the tire would most likely involve a large amount of muscle strain. The center of gravity of the body, plus the jack and the power tool, are extended far beyond that occurring in the spare tire load operation. As noted, the simulation of the jack load operation was estimated rather than actual as with the spare tire load operation.

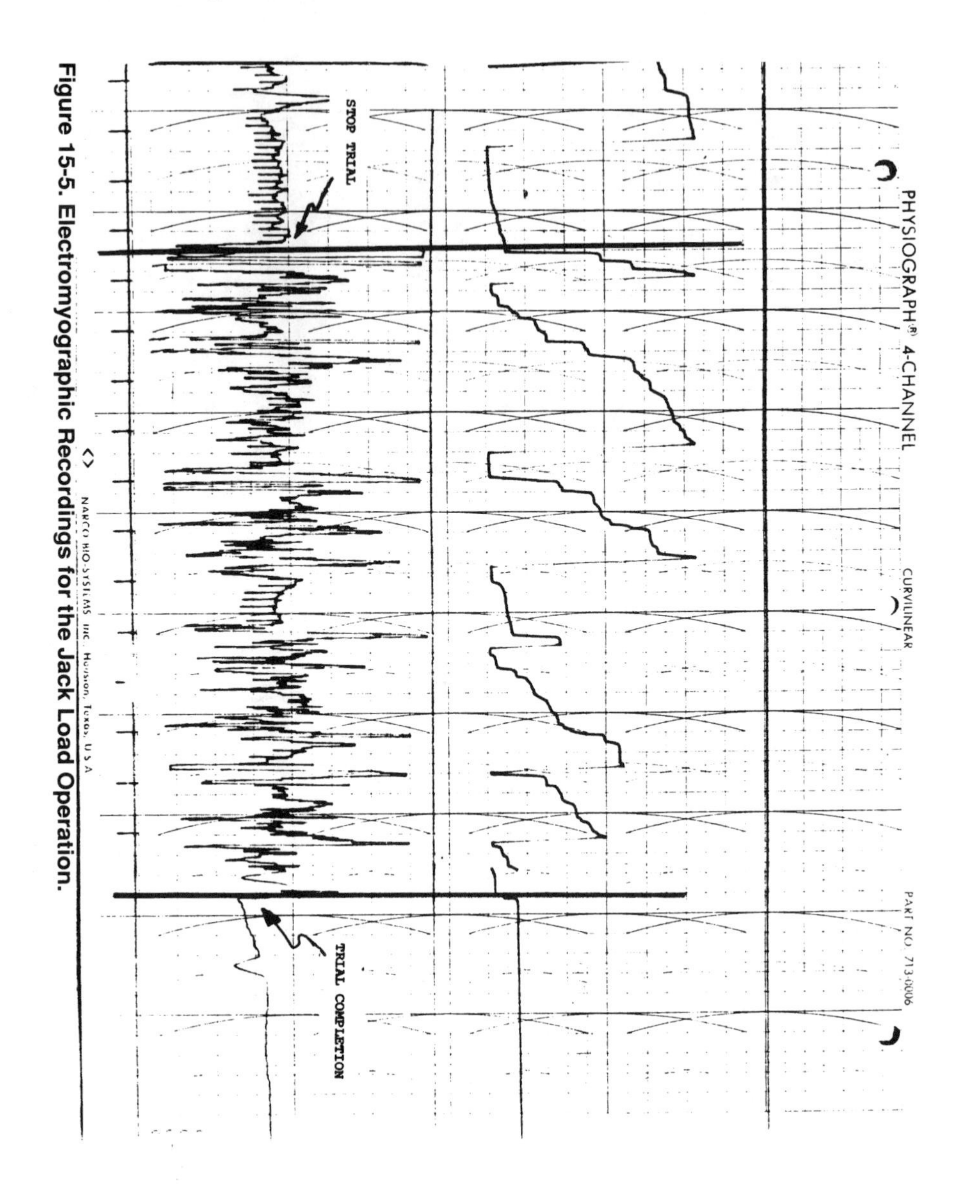

Figure 15-5. Electromyographic Recordings for the Jack Load Operation.

RECOMMENDATIONS

Spare Tire Load Operation

When a strenuous job is encountered, the tendency is to place a larger, stronger person on that job. The results of the analyses indicate, however, that a taller, heavier person would be at a disadvantage, because the increased torso weight and the distance from the L5/S1 to the torso's center of gravity when in a bent posture would actually result in more severe back strain. Selection of the worker, therefore, does not appear to be a particularly good solution to this problem.

Another important characteristic of this job is that the workers assembled the J-hook, jack plate, and wing nut while bending over into the trunk. By using a longer tie-down bolt and a larger "eye" in the trunk and preassembling these components prior to leaning over into the trunk, the workers were able to reduce the time in the stress-related position.

The investigators recommended the following simple modifications to the assembly procedure to reduce the stress to the back:

1. Preassemble tie-down components, outside of the trunk
2. Lower the tire with the two-hand distributed method
3. Increase the size of the "eye" into which the J-hook fits
4. Tighten the wing nut by reaching in from the side of the rear quarter (extending the power tool)

Jack Load

Apparently, no specific recommendation can be made based upon the results of this investigation other than that the rear installation from behind the car appears to be undesirable. This conclusion is supported by the fact that the spare tire load task has a history of medical problems and worker dissatisfaction; and the rear installation task appeared to be even more severe.

References

1. Snook SH. The design of manual handling tasks. *Ergonomics*. 1978;21:12;963-986.

2. *A Work Practices Guide for Manual Lifting*. Cincinnati, OH: US Dept. of Health and Human Services, National Institute for Occupational Safety and Health; 1981; Technical Report No. 81-122.

3. Habes D. *Muscle Activity of the Low Back Associated with Repetitive Learning Tasks in Industry*. Ann Arbor, MI: University of Michigan; 1976. Thesis.

Steven Johnson is President of Ergonomics Analysis, Inc., a consulting firm specializing in applications of ergonomics in industry. He is also professor of industrial engineering at the University of Arkansas.

Dr. Johnson is a registered professional engineer and a certified quality engineer. He earned BS and MS degrees in Engineering Psychology from the University of South Dakota and the University of Illinois, respectively.

Dr. Johnson is a former director of the Ergonomics Division of the Institute of Industrial Engineers where he is also a member. He is also Chairman of the Industrial Ergonomics Group of the Human Factors Society.

16

Making a Difference at Johnson & Johnson: Some Ergonomic Intervention Case Studies

Arthur R. Longmate
Timothy J. Hayes

Industrial ergonomics is no place for heroes. Individuals who can make it *all* happen simply do not exist. Many different functions, such as medical, engineering, safety, management, and staff, must continually interact to resolve ergonomic issues. Why is all this interaction necessary? The answer lies within the complex issues of many ergonomic problems. To investigate and document a problem is generally not difficult, but to determine an effective solution is often very hard. Truly effective solutions are born out of cooperative efforts, in which outcomes are shaped by expertise and practical experiences from a wide variety of sources.

Ergonomic initiatives can take many forms, such as engineering design and/or redesign, ergonomic training, comprehensive medical case management, and on-the-job exercise programs. These initiatives primarily focus on solving ergonomic problems, which can lead to cumulative trauma disorders (CTDs) and materials handling injuries, two of the most common medical conditions caused by work-related activities.

Ergonomic efforts primarily focus on two areas: *prevention* and *intervention*. An example of problem prevention is the ergonomist who is invited to participate in the design of a new process at the concept phase and is an active member of the project team. Incorporating ergonomic design during the early stages of a project is extremely cost effective when compared with potential costs after the project is completed. At Ethicon, Inc. (one of Johnson & Johnson's largest domestic companies) this type of cooperation and interaction is required, with ergonomic considerations integrated early, before a new manufacturing process or machine is implemented.

The goal of ergonomic intervention is to resolve an existing problem. This activity usually falls into one of two categories: job redesign or medical case follow up. During the job redesign process, the ergonomist systematically reviews existing operations, stressful work practices, and recommends ways to correct the problem. The medical case follow-up process is essentially the same, except that the ergonomist works closely with the injured worker to facilitate a quick recovery and to recommend ergonomic solutions that will prevent a recurrence.

In the capacity of problem preventer and solver, the ergonomist performs a wide variety of duties, including:

- Review current medical incidents
- Develop a medical incidence rate reporting system to determine priorities and evaluate progress
- Assess existing jobs to identify high-risk work practices
- Develop recommendations to eliminate or alleviate these work practices
- Develop specific redesign projects and, unless other resources can be identified, determine costs, sell projects to management, oversee projects through completion, and evaluate results
- Interact in new equipment design projects from concept through implementation
- Develop simple, concise, and practical ergonomic guidelines for engineers' and managers' day-to-day use
- Develop and conduct ergonomic training programs for workers, managers, and engineers

A full-time, dedicated ergonomist is essential to a successful industrial ergonomics program. Once again, the program's overall effectiveness depends on how well resources from other disciplines interact to comprehensively address ergonomic issues. The following case studies illustrate how engineering and ergonomic solutions helped solve several ergonomic problems.

Case Study One

Large Bandage Making Machine

The large bandage-making machine combines wide layers of raw material and cuts them into the final shape. Each layer is fed into the machine from separate raw material feed rolls. The machine's operator must replenish the rolls approximately six times per day. While no injuries have been reported on this job, a machine operator suggested that the unwind stand for the largest raw material roll be modified to reduce the risk of back strain. Supporting this request was the fact that two female operators had recently attempted the job and could perform all aspects except for changing the one large material roll.

The staff ergonomist used the National Institute for Occupation Safety and Health (NIOSH) *Work Practices Guide for Manual Lifting*[1] to analyze the job elements required to handle the material rolls. The results, shown in Table 16-1, verify that a problem exists, particularly when the operator positions a new roll of material onto the unwind stand. An extreme horizontal (H) distance is required for two reasons:

1. The roll's diameter measures 18 in. The operator lifts the roll by using a mandrel inserted through the core, with the roll's axis parallel to the front of the operator's body (Figure 16-1). This lifting caused excessive stress on the back.
2. The unwind stand base was built with a piece of angle iron running across the floor at the point where the operator stands to position the roll. This interference substantially increases the required H distance.

Based on the machine operator's suggestion and the ergonomist's analysis of the job requirements, management agreed to redesign the unwind stand and mandrel. The new design allows operators to position the material roll on the side rather than on the end of the stand (Figure 16-2). One feature of this new design is the permanent attachment of the mandrel to the unwind stand, which reduces the lifting requirement by approximately 35% (19 lb). Another feature eliminates two severe finger pinching points where the mandrel slid into the slots in the original mandrel design (Figure 16-1).

The unwind stand's new design also allows operators to lift the roll with the flat side of the roll against their bodies, which reduces the required H distance to 10 in (Table 16-1). Because the allowable weight limit is increased from 25 lb to 51 lb, a greater number of workers are now able to safely perform this job.

A key factor in selling this engineering change to management was the ergonomist's use of the NIOSH *Work Practices Guide for Manual Lifting* to demonstrate the reduction in physical stress gained through the proposed modification. Perhaps the greatest use of the NIOSH *guide* and other models is to demonstrate relative improvements, which can be achieved through equipment redesign, before the change is actually implemented. These modifications were favorably received by the workers, especially the two women who are now able to rotate into this job without apparent difficulty.

Table 16-1. Large Bandage-making Machine NIOSH Lifting Guide Analysis.

	Lift Distances (inches)					
Job Design	H	V	D	Frequency (Lifts/min)	Roll Weight (lb)	Action Limit (lb)
1A. Original	20	25	5	.013	54*	25
1B. Modified	10	25	5	.013	35**	51

H = Horizontal; V = Vertical; D = Distance moved
* Including mandrel
** Roll only

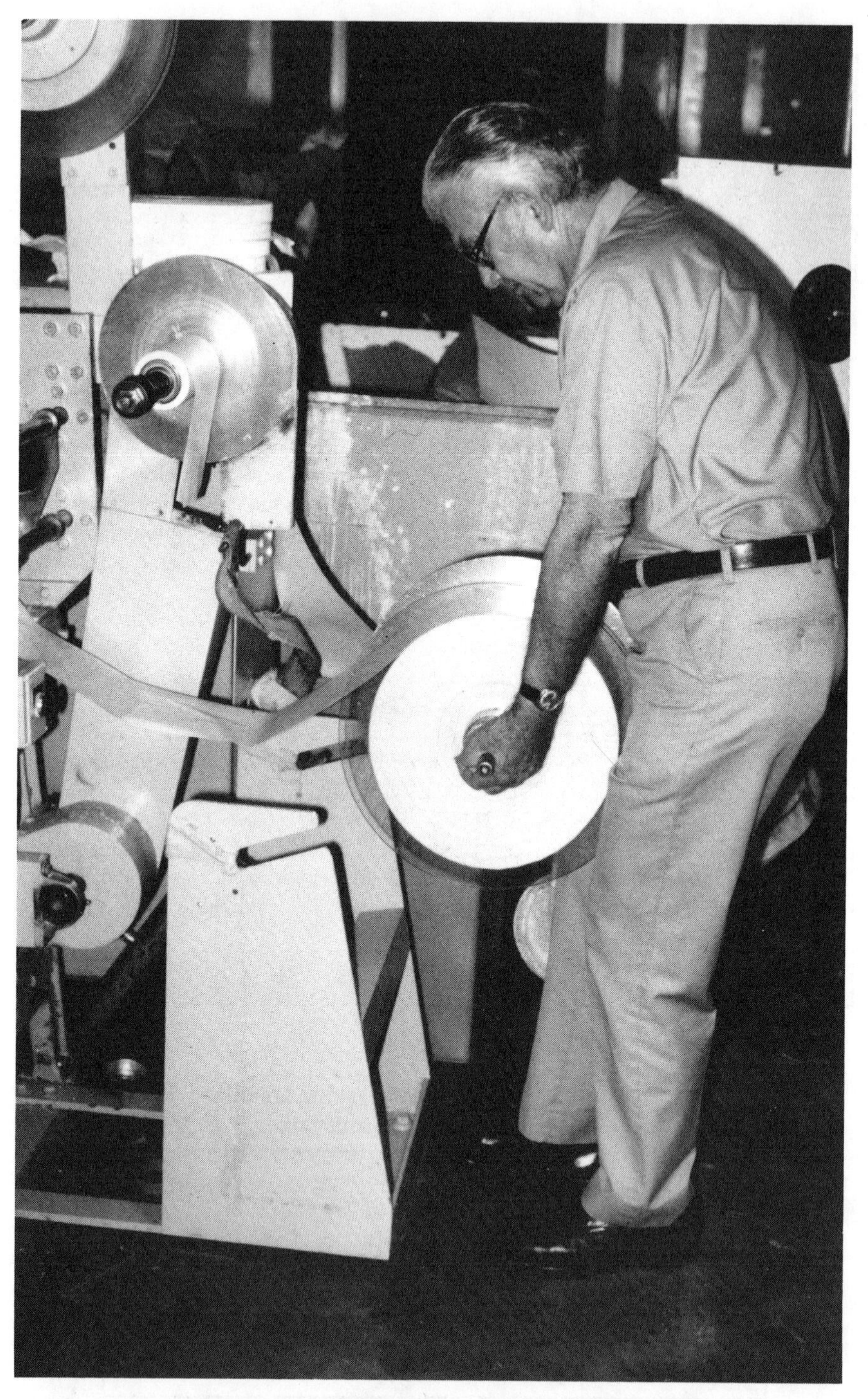

Figure 16-1. Large Bandage Making Machine Before Modification.

Figure 16-2. Large Bandage Making Machine Suggested Modification.

Case Study Two

Semi-Automatic Winding Machine

The semi-automatic winding (SAW) machine winds sutures in a figure 8 pattern and then folds a paper dispenser around the suture. The machine is a significant labor-saving device. Prior to the purchase of the SAW machine, workers in the hand winding department experienced a high incidence of cumulative trauma disorders of the wrist and hand due to repetitive motions. The SAW machine effectively eliminated the repetitive winding motions, and CTDs of the wrist and hand are now rare. Several unanticipated problems did develop after the introduction of the machine, however.

The SAW machine operator sits at the machine and feeds sutures into the loading area. The sutures are picked up by the machine, wound, and then placed in a folder. On average, the operator will load approximately 10,000 sutures during a shift. This number can vary depending on suture type and length. If the machine jams or if some other event occurs that stops the machine, the operator will clear the jam and then hit a foot pedal to restart the machine. If there are machine or product problems, the operator may hit the foot pedal up to 400 times per shift.

The working height of the machine is 37 in. Leg room varies from 11 in to 14 in, and a cut-out is provided for the operator's feet. The foot pedal is mounted 7 in above the floor on the plate located in the cut-out.

Workers frequently complained about the machine. Large operators complained of the limited leg room, small operators said they had trouble reaching the foot pedal, and all complained about the sharp edges of the work table. A number of injuries were also noted, some of which were reported to the Occupational Safety and Health Administration (OSHA). One case of sciatic nerve entrapment prompted a worker's compensation claim. The injured worker was about 5 ft tall and had to sit on the edge of her chair in order to reach the foot pedal (Figure 16-3).

In response to workers' concerns, management agreed to make the following modifications to the SAW machine:

- Enlarge the cut-out in the front plate to provide more foot room and to accommodate an adjustable foot pedal
- Modify the pedal to adjust between 7 in and 15 in (Figure 16-4)
- Install padding on the edge of the work table to keep the sharp edges from cutting into the operators' forearms

Increasing the leg room depth was also investigated but later dropped, because it could not be accomplished without an almost total redesign of the machine.

At 7 in above the floor, less than 5% of the operators could reach the stationary foot pedal while sitting in a normal position. As a result of adding the adjustable pedal, 95% of the operators can now reach the pedal without having to sit on the edge of their chairs. The padding has also been a popular change, and complaints about the work table digging into the operators' forearms have stopped. To date, the changes have only been made in one plant, but are scheduled to be implemented at all locations.

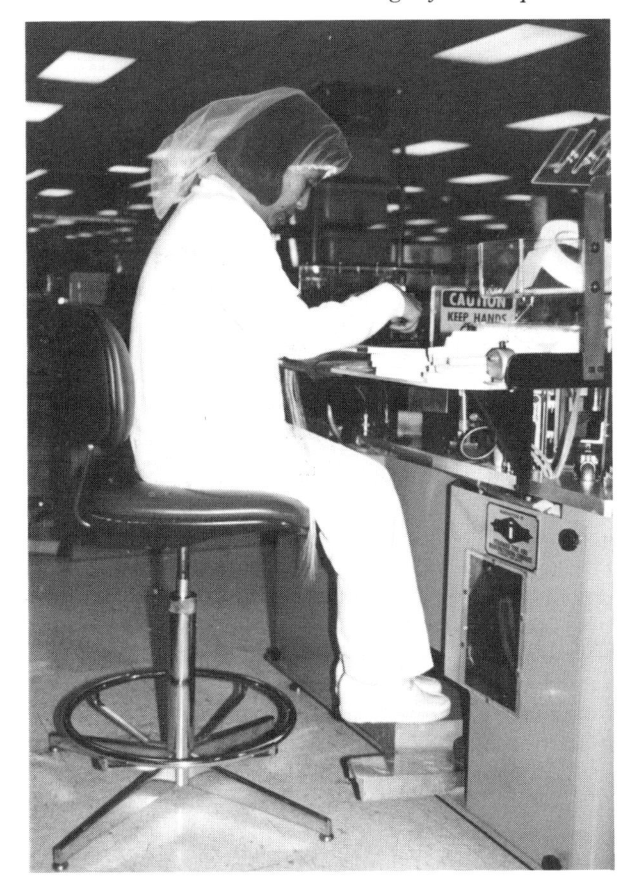

Figure 16-3. SAW Machine Before Modification.

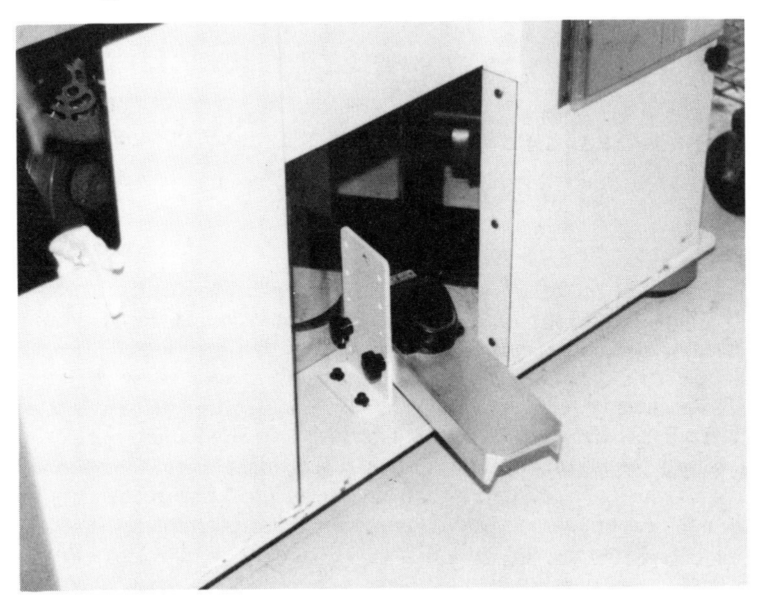

Figure 16-4. SAW Machine Adjustable Foot Pedal.

Case Study Three

Stapler Assembly/Weld Line

Disposable, sterile skin staplers are relatively simple, low-cost devices, which are used for both external skin closure and internal surgical procedures. They are produced in high volumes compared with some of the more complex internal medical stapling devices.

ASSEMBLY OPERATION

Workers assemble skin stapler components on conventional flat tables (Figure 16-5). The five component parts are delivered to the tables in rectangular tote pans, measuring 12 in x 19 in x 6 in. Assemblers use two different methods to pick up the component parts: reaching into the tote pan or scooping handfuls of parts out of the pan and onto the table.

Reaching into the tote pan each time resulted in a repetitive wrist flexion and/or ulnar deviation, particularly when the pans were less than one-half full and assemblers had to reach near the bottom of the pan. Scooping handfuls of parts out of the pan onto the table, on the other hand, resulted in concentrations of high mechanical forces to the hands and fingers. Also, digging the parts out of the pan required high hand force, and parts tended to become spread out over a large area and mixed up with other components. Consequently, assemblers had longer and more difficult reaches to grasp the necessary parts. Emptying the parts onto the table was probably the best method, but it caused an extreme intermixing of parts and generally congested the work area.

Most workers assembled the five components by using the nondominant hand to hold the stapler's handle, while using the dominant hand to assemble the remaining components onto the handle (Figure 16-6). After the stapler was assembled, the assemblers placed it into a tray. When the tray was full, the assembler pushed it along the table top to the final welding station. From an efficiency viewpoint, using the nondominant hand as a fixture is ineffective. From an ergonomics viewpoint, however, this method was preferred as it minimized awkward extremity posture during assembly.

WELDING OPERATIONS

The welder used an ultrasonic welder to weld the assembled stapler. The welder then fired the stapler five times to test staple formation and staple feed into the magazine (Figure 16-7). The job steps follow:

1. Use the left hand to pick up the assembled stapler from the tray.
2. Use the left hand to position and insert the stapler into the welder nest.
3. Use the left hand to close the manual clamp on the welder nest and to secure the stapler into the nest.
4. To operate the welder, push and hold the welder activation buttons.
5. Use the right hand to remove the stapler from the welder nest.
6. To break the stapler free from any weld flash that might have formed, use the palm of the left hand to fire the stapler once by forcefully striking the trigger.
7. Fire the stapler four additional times to observe staple formation and proper staple feed into the magazine.

8. If acceptable, place the stapler into the final cleaning station. If unacceptable, destroy the stapler.
9. Record any defects on the sheet.

Using a four-man crew (two assemblers, one welder, and one cleaner/packer), the production rate on this line was approximately 4,000 instruments per day. Job rotation was not structured, and generally the fastest person ran the welder. As the welder station became overloaded, one of the assemblers would swing over and assist the welder by performing the test-firing function. In this case, the welder would only perform the welding operation, and hand the instrument to the temporary helper for test firing. This swing arrangement seemed to be the most efficient way to deal with the inherent line imbalance between the assembly and welder stations. One assembler could not work fast enough to keep the welder continuously supplied, and two assemblers could produce enough assemblies to eventually fill the queuing space.

In addition to the line balancing and component picking problems, the ergonomic-related medical incidence rate in this department was extremely high. The main problems included various forms of tendonitis and other hand/wrist-related disorders. Many workers were placed on medical restrictions, which created job-manning problems for department supervisors since few restricted work load jobs were available.

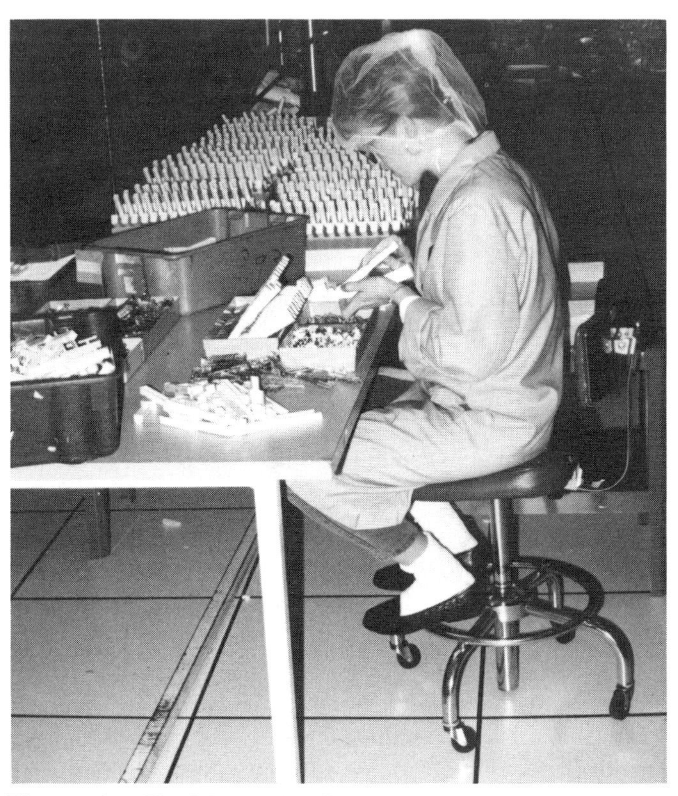

Figure 16-5. Final Assembly Before Modifications.

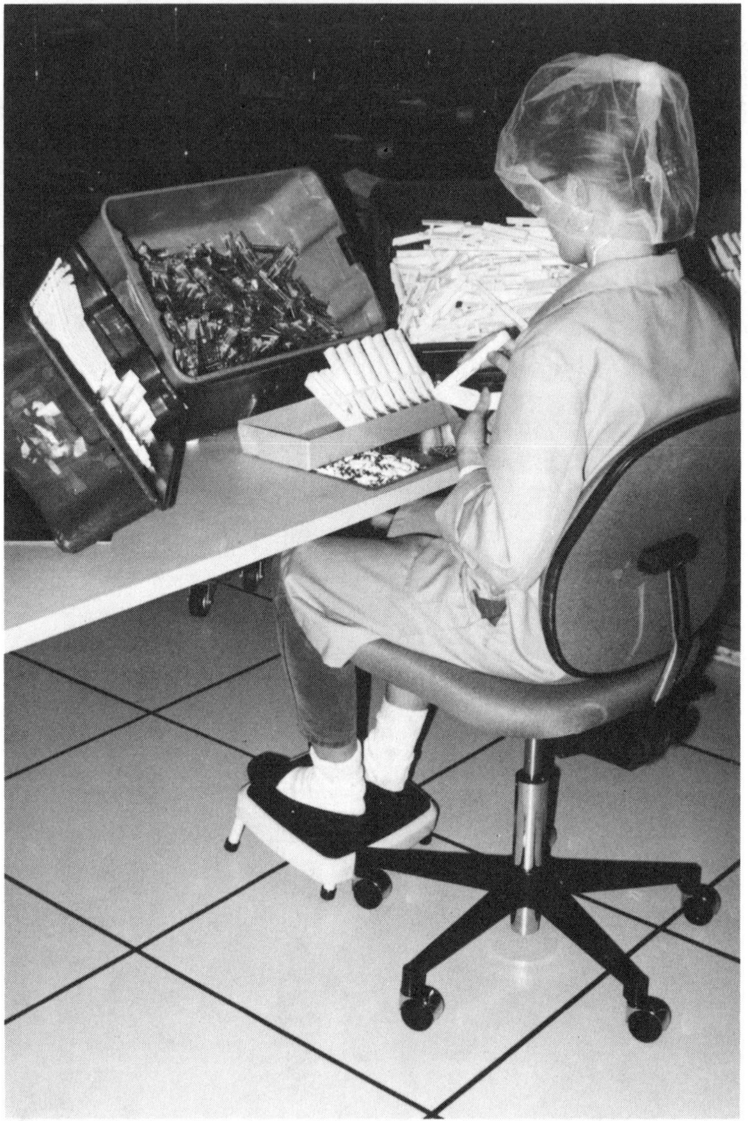

Figure 16-6. Final Assembly Modified Station.

Work Station Modifications

The ergonomist assessed the problem, and management agreed to implement the following modifications:

- Initiate a structured, job rotation sequence, during which workers rotate job positions every 30 minutes.
- Provide the assembly stations with adjustable V-stands to tilt the tote pans

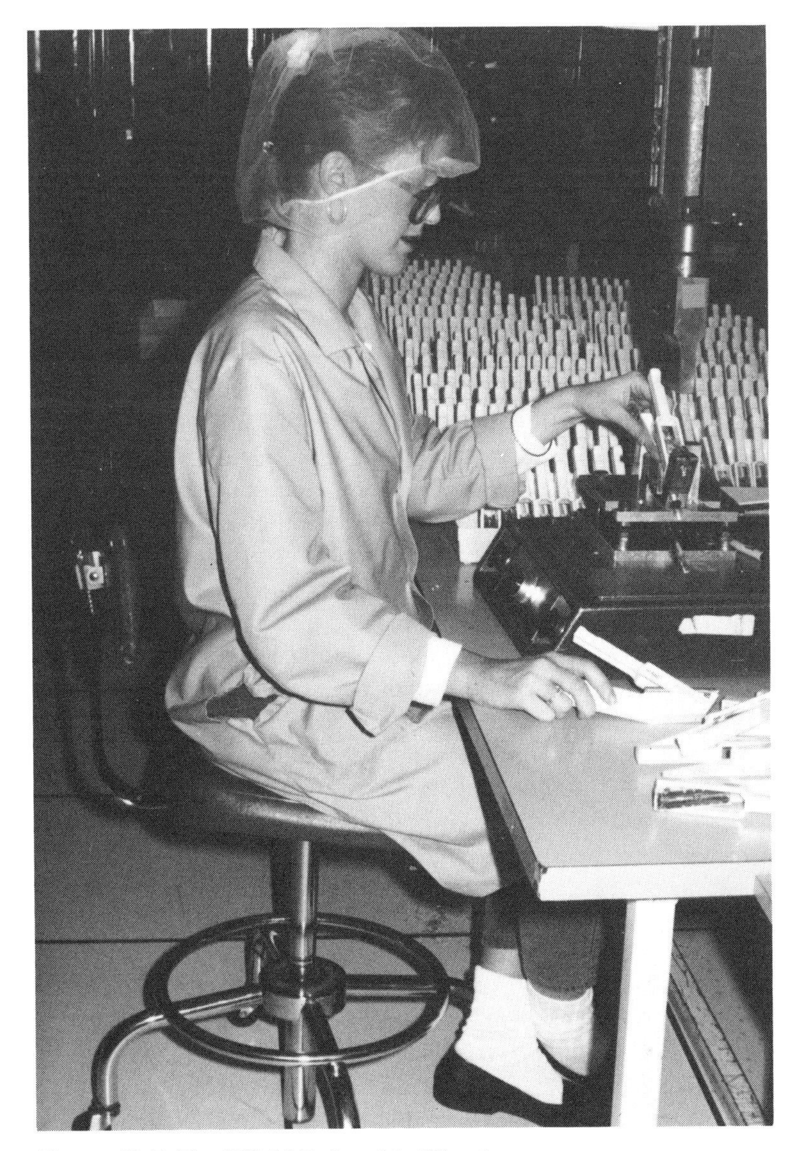

Figure 16-7. Final Weld Before Modification.

to a more accessible angle. Also, for two small parts, provide recessed parts trays positioned directly in front of the assemblers. This modification allows assemblers to easily grasp the components, with minimal wrist bending. To replenish the recessed part trays, provide a metal scoop so assemblers can scoop parts from the tote pans.

- Purchase adjustable ergonomic chairs for all workstations. From the seated position, assemblers can now easily adjust the seat height, inclination, back angle, and back height (Figure 16-8). Previously, assemblers had their own personal chairs, which were adjusted to their preference. When they rotated to new positions, the working height of the chair was different and could not be adjusted quickly. The new chairs eliminated the need to move individual chairs from station to station.

- Provide footrests for shorter assemblers. The long-term goal is to provide adjustable footrests at each workstation, which can be adjusted without crawling under the workstation. Several types of adjustable footrests are available, but none have been identified with sufficient height range and easy adjustment from the seated position. The interim solution is to provide simple footstools on which the legs are cut off to provide several different heights. The legs on one side are cut off about 1 in lower to provide a slight tilt of about 10°. This temporary solution still requires assemblers to take their footstools around with them as they rotate from position to position (Figure 16-6). Design of this easily adjustable footrest is proceeding in-house, and will be built by an outside contractor.
- Implement a presence-sensing activation button system to operate the ultrasonic welder (Figure 16-8). Of all the options considered, this system is the most functional and cost effective. The diameter of target contact area on the buttons, for example, is 3 in as opposed to 1 in. The new design requires no activation force as both hands just simply have to contact the buttons. Also, an adjustable angle mounting bracket attached the activation buttons to the sides, rather than being mounted on the front and base of the welder. The angle of the buttons easily tilts up to approximately 20° toward the welder. In the old design, the buttons were at a steep angle (60° - 70°), which required substantial activation force. The original button separation was only 12 in, center to center. The new separation is 18 in, which is safer because it more closely approximates normal shoulder separation. These modifications have reduced the high incidence of thumb tendonitis caused by the old button placement and size.
- Integrate a pneumatic clamp, to automatically clamp the stapler to the weld nest. This eliminates the repetitive striking of a manual De-sta-co clamp to hold the stapler during welding. The manual clamping method also involved a rotation of the forearm in excess of 90°, while sustaining a fairly long reach. This combination of risk factors significantly contributed to a variety of hand/wrist, elbow, and shoulder problems.

The final engineering modification, which is still under consideration, is to provide a conveyor to automatically transport trays of assembled instruments between the assembly and welding workstations. Currently, the assemblers lift trays of assembled staplers off of the table and place them onto a series of connected tables, which go to the welder. The assemblers continually push the trays along on the table top to the welder, which forms the queue of trays. The welder must get up occasionally to pull trays over to the welding station if the trays are not pushed into position. This arrangement is clearly inefficient. With

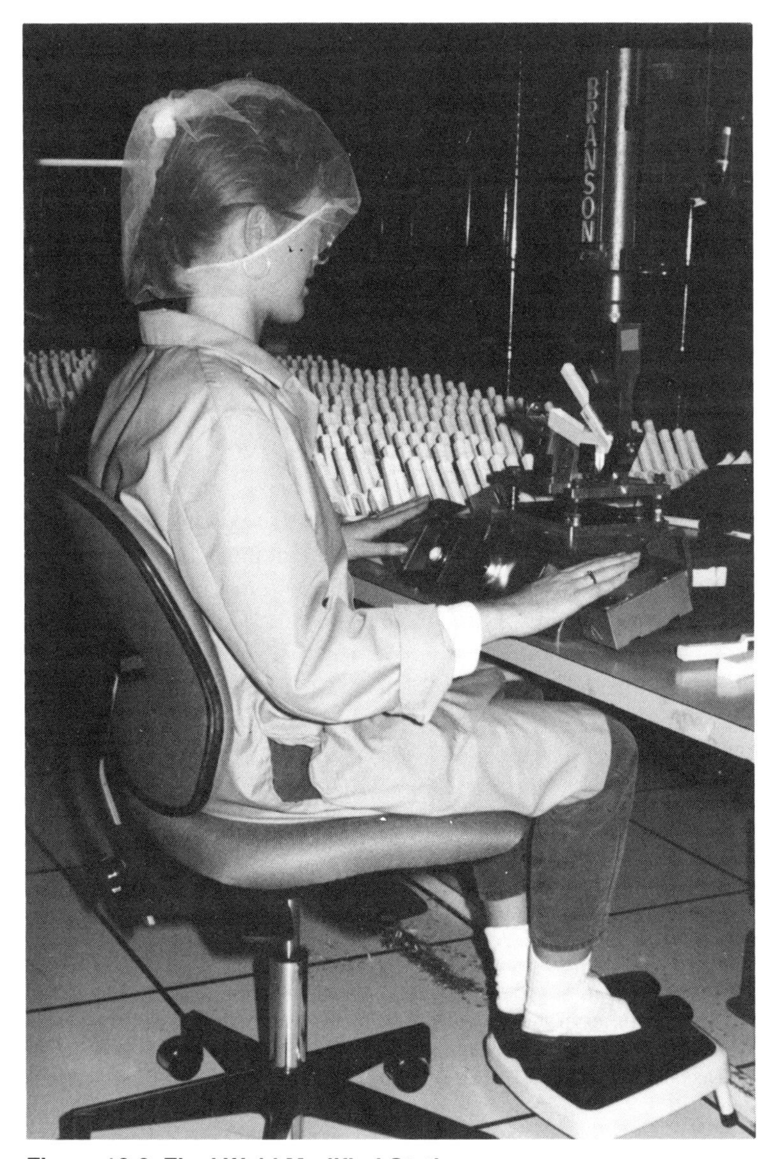

Figure 16-8. Final Weld Modified Station.

the conveyor, the assemblers would simply slide the trays over the table directly onto the conveyor. As the welder removes an empty tray from the end position, a full tray would automatically be pushed into position.

Ergonomically, the benefits of a conveyor are not clear, although eliminating the tray handling at the assembler station is desirable. The full trays of staples only weigh approximately 5 lb, however, so there certainly is no critical biomechanical problem. In fact, simply performing this occasional tray handling task may provide a short break from the tedious and repetitive assembly process. At the welder station, the task of getting up to retrieve a new tray of staplers, although inefficient, is probably beneficial from an ergonomic standpoint. It provides a short break from the tedious and repetitive welding process.

If not for the new 30-minute job rotation process, the conveyor project would probably be pursued based upon ergonomic justification. In light of the reduction of repetitivity through job rotation and considering the obvious productivity improvements, the conveyor project is still under consideration. Due to the relatively high cost of the conveyor, approximately $8,000 per line, and due to the questionable ergonomic impact, justification remains in question.

Figures 16-6 and 16-8 show the modified final assembly and final weld workstations, with the exception of the conveyor between stations. Worker response to the modifications has been extremely positive, and management has noted an increase of productivity of 10% to 12%. Reports of new work-related medical problems have greatly diminished. Unfortunately a considerable number of workers in the department still have existing medical problems, which developed prior to the modifications. Plant medical personnel continually work with these workers on programs of rest from repetitive work, physical therapy, and medications. Through this combined approach of engineering process changes and comprehensive medical case management, medical incidence rates in this department are expected to continue to decline.

Reference

1. National Institute of Occupational Safety and Health. *A Work Practices Guide for Manual Lifting.* Cincinnati, OH: Department of Health and Human Services; 1981. Technical Report No. 81-22.

Arthur R. Longmate received a BSIE, a MSIE in Occupational Safety and Health Engineering, and achieved Candidacy in the PhD program at the University of Michigan. Mr. Longmate is currently a Staff Industrial/Ergonomics Engineer at Ethicon, Inc., a Johnson & Johnson company. He is a senior member of the Institute of Industrial Engineers, the Human Factors Society, and the American Industrial Hygiene Association. He has also served for five years on the Presidents Committee for Employment of the Handicapped.

Timothy J. Hayes received a BS in Psychology and MS in Industrial Engineering from the University of Wisconsin-Madison. He has been employed as an ergonomics engineer at Ethicon, Inc. in Somerville, New Jersey for three years. Previously, he worked as a human factors engineer at FMC in San Jose, California. Mr. Hayes is a member of the Institute of Industrial Engineers and the Human Factors Society.

17

A Perspective on Solving Ergonomic Problems

Daniel J. Ortiz
Susan M. Gleaves

The US Department of Labor grants federal money to each state to provide in-state employers with confidential safety and health technical assistance at their request and at no cost. These state programs, which are independent of any compliance or enforcement activity, can either provide ergonomic assistance or refer companies to consultants who have the required expertise.

Through detailed work place evaluations, the Georgia Institute of Technology Environment, Health, and Safety Division assists companies in Georgia with increasing compatibility between workers and their respective environments. The goal of this program is to not only recommend workplace improvements, but to educate companies to ergonomically view and assess their facilities on a continuous basis.

APPROACH

Most often companies request technical assistance after the fact (intervention) in response to a high prevalence or the sudden appearance of cumulative trauma disorders (CTDs), such as carpal tunnel syndrome, tendonitis, and tenosynovitis; back injuries; and/or performance-related problems. Companies also occasionally request an ergonomic evaluation to prevent a problem from occurring and to improve efficiency.

On-site, the evaluation process begins with a review of the injury and illness records and the company's medical surveillance system. This evaluation helps identify the jobs or locations in which the prevalence of ergonomic-related disorders are higher than expected. Ergonomists then calculate incidence rates and make comparisons between departments and jobs. Those jobs where problems exist are then targeted for further review. Also, information on worker complaints, turn-over, and production-related problems is collected from management, when it is available.

Once the problem jobs are identified, specific workstation and worker measurements are taken. The height of the work, location, weight of materials, push/pull forces, and job cycle time, for example, are particularly important for evaluating postural stress and work tolerance. Likewise, anthropometric data concerning worker stature, elbow height, and reach envelope are recorded. The gender, age, grip strength, and experience level of the target workforce are also obtained when necessary.

SUBJECTIVE ASSESSMENT

A structured interview and/or administration of a comfort questionnaire is an excellent method to get direct feedback from the individual workers about their jobs. Workers know their jobs best and can provide excellent information about job demands and what is needed to reduce the stress. Many methods for obtaining worker input have been developed. The method used must be easy to administer, require little time to complete, and be unobtrusive, especially where workers are on an incentive, piece work payment system, or where the work is machine paced. Injury and illness data and employee input, taken together, provide valuable information about workstation and methods design and also provide a baseline for comparison should any changes be implemented.

Electromyography (EMG), which measures the activity of selected muscle groups, is sometimes used for comparing job methods and the force requirements of different jobs or job components. Heart rate data are also sometimes collected to provide input into the physiologic cost of task performance.

ENVIRONMENTAL CONSIDERATIONS

Lighting is an important environmental consideration, because a poorly lit workplace can impact job performance by causing workers to assume stressful postures to see their work. Evaluation of work station lighting involves taking *illuminance* (the amount of light falling on a surface) and *luminance* (the amount of light reflected from a surface) readings and comparing them with the Illuminating Engineering Society of North America's recommended values.[1]

The workplace is also carefully examined for sources of potentially debilitating direct and indirect glare. Fine assembly, such as printed circuit board component insertion, sewing tasks, and inspection tasks, are examples of visually demanding jobs. Other important environmental factors, which often need to be evaluated, include the thermal conditions, sound level, and air quality.

Video Observation

To evaluate posture and job techniques, representative workers performing target tasks are video taped from usually, the right, left, and back sides for at least one job cycle. If possible, workers with at least one year experience and with less than six months experience are taped. This is crucial with labor intensive operations such as a cut-up line in poultry processing, in which several people might use different techniques to make the same or a similar cut with a knife. Typically, the inexperienced worker uses inefficient, forceful motions to accomplish the same job as the experienced worker who uses fluid, smooth motions. Any observed differences in job technique can be used as part of a training program to encourage workers to adopt biomechanically sound motions and postures.

JOB ANALYSIS

All the information collected in the field is analyzed in a laboratory. Material handling tasks are analyzed using a two dimensional biomechanical model, developed by the University of Michigan Center for Ergonomics[2]; the NIOSH *Work Practices Guide for Manual Lifting*[3]; and/or psychophysical charts, developed by Stover Snook, PhD, Liberty Mutual Insurance Company.[4] Each method has its limitations and special applications. For the most part, however, the analytical approaches assume ideal lifting conditions, such as smooth lift in front of the body, and so the lifting or material handling limits derived probably underestimate the risk of back injury. Consequently, the video tape supplies valuable information about the posture and motions that could actually multiply the load effect on the back. Bending, twisting, and frequent forward reaches beyond 16 in are examples of such high-risk work practices.

Video Registration and Analysis

A computer method known as video registration and analysis (VIRA) effectively analyzes repetitive jobs. The video taped jobs are viewed in slow motion, and target postures are categorized according to parts of the body and movement angle. Using the computer as a stopwatch, the percent and duration of the cycle time the worker spends in each posture category and the number of posture changes can be documented; for example shoulder abduction (elbows raised away from the body), wrist deviation and flexion, torso bending, and arm extension. The following illustrates how this method is applied.

A machine forms and ejects finished components to a packing station. The packer retrieves 10 parts and places them in a bag. The bag is then sealed and placed in a box for shipping. This cycle is completed in 30 seconds and repeated through a 12-hour work shift. In response to complaints of low back pain by the majority of the packers, this job was evaluated.

Video registration and analysis revealed that workers spent 20% of the time in a stooped posture (back flexed greater than 45°) making six posture changes per minute (going from upright to stooped) in order to acquire and pack the components. This translated to 2,880 posture changes per work shift and a huge lifting requirement for the lower back. In response to this analysis, the work station was redesigned to allow workers to assume a more upright posture to accomplish the task (Figure 17-1 and Table 17-1).

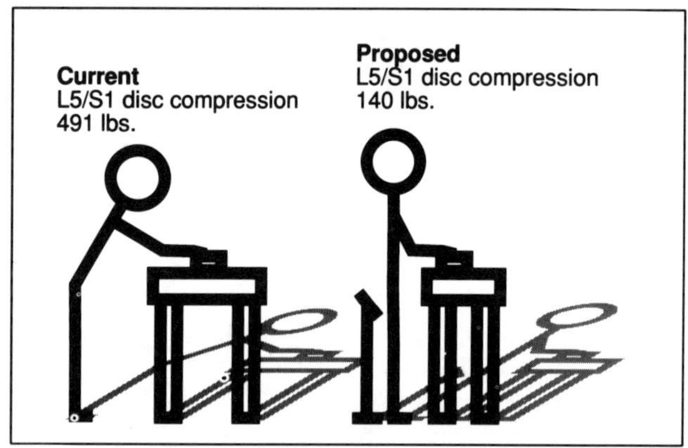

Figure 17-1. Packing Work Station Illustration.

Classic safety and health workplace considerations should not be overlooked as work stations or processes are designed or modified to relieve postural stress. For example, if a work bench is situated higher than it should be from an

Table 17-1. Ergonomic Considerations.

Stressor	Target	Solution
1. Forward reach > 16 in	Arms/shoulders/back	Modify workstation so work can be done with elbow down and forearm flexor angle approximately 90°.
2. Forward stooping	Back	Modify workstation and adjust machine pace to eliminate excessive reaching to "stay ahead."
3. Standing	Legs	Sit/stand stool and/or anti-stress mat.

ergonomic standpoint, creating shoulder and arm discomfort, and below the workbench is an exposed motor V-belt drive, it is not enough to lower the table to reduce or eliminate the biomechanical stress. Obviously, the machine hazard must be guarded to prevent a serious injury from occurring. Safety, health, and ergonomics are inextricably entwined, so examining the total picture, not just one aspect, to increase compatibility at the worker-workplace interface is important.

CASE STUDIES

We have conducted evaluations for numerous companies, representing a wide variety of industries. The observations and findings generally agree with what is accepted in the scientific and engineering communities. Cumulative trauma disorders occur with jobs that require repetitive movement of the hands, wrists, and arms, and back injuries are associated with material handling intensive operations. With this in mind, the following composite case studies describe ergonomic stressors and corresponding control strategies, which we have found in certain poultry processing, apparel manufacturing, and electronics assembly operations.

Poultry Processing

In the poultry processing plants, CTDs are common, particularly carpal tunnel syndrome and tendonitis. As expected, the incidence of CTDs is greatest in hand-intensive jobs, such as those that use knives and scissors. Deboning and cut-up jobs are cases in point.

The primary ergonomic problems noticed in poultry processing include:

- Many of the cut-up tasks require workers to make at least 4,800 cuts each day with knives or scissors
- Highly repetitive work is frequently done with a flexed or deviated wrist
- Processing lines are often too high for shorter workers
- Off-line trim and salvage tables are often too low
- Workers typically stand for extended periods

Ergonomic intervention strategies included line height adjustment, raising the shorter workers with the use of platforms, experimentation with biomechanically designed tools, and the introduction of anti-stress mats. Many companies are also implementing detailed training programs which include safe and ergonomically sound job methods, job rotation, knife sharpening, and early detection programs.

Apparel Manufacturing Sewing

Carpal tunnel syndrome and tendonitis are prevalent in some apparel manufacturing sewing operations, and back and neck pains are often reported on comfort questionnaires. Sewing operations frequently require intensive hand movement to manipulate, support, and guide the fabric to and from the sewing machine. The following are primary problems we have observed in apparel manufacturing sewing operations:

- Some highly repetitive sewing tasks involve more than 20,000 wrist posture changes in an eight-hour work day
- Reaches to both sides involving bending and lifting are frequently required to acquire fabric bundles and move apparel
- When the chair and machine heights are not adjustable, workers often have a less than optimal work height
- Chairs rarely supply much-needed lumbar support and frequently have home-made padding

Ergonomic intervention strategies encountered in sewing operations include the introduction of padded, adjustable chairs with lumbar support and swivel capability; modified parts ranging from bundle size limitations to spring loaded carts; tilt table and height adjustable work tables both support and space for the feet and legs; and the introduction of biomechanically designed tools, including spring loaded scissors, and automatic clippers. Training, build-up periods, and conservative treatment plans are other methods used in some plants to help reduce the risk of injury.

Printed Circuit Board Assembly

Cumulative trauma disorders, including tenosynovitis, tendonitis, and carpal tunnel syndrome, backaches, and headaches are often associated with some printed circuit board assembly tasks. Circuit board assembly tasks require workers to grip and regrip small parts, manually position parts, and assemble items, using a great deal of repetitive wrist motion with forceful exertion. Common problems include:

- Location of materials and parts bins at the floor level and/or outside the workers' normal reach envelope requires excessive bending and reaching
- Excessive reaches are often complicated by improper tool or task orientation
- Workers often raise the printed circuit board to eye level, and support it with their nondominant hand. At the same time, the dominant hand uses dikes and/or pliers to insert small components with a bent or deviated wrist
- Poor ventilation and poor lighting are often associated with headache complaints

Ergonomics applications in printed circuit board assembly operation include automating some of the highly repetitive tasks, using clamp fixtures to support the circuit board; altering the shapes and locations of supply bins; using properly designed chairs; and introducing biomechanically designed hand tools as well as counterbalanced, overhead suspensions for heavier tools. Ventilation and lighting problems have been corrected by installing local exhaust ventilation systems and by changing the locations, numbers, wattages, and types of lighting fixtures. Increasing task cycle lengths and introducing job rotation are also common control strategies in printed circuit board assembly operations.

CONTROL STRATEGIES

After visiting a company and assessing how ergonomic intervention could improve the workplace or reduce the amount of stress on workers, specific ergonomics applications are recommended. For employers the initial cost of some options can be very high and very time consuming, but can often be justified by reduced workers compensation losses. Other options require little or no initial implementation expense. All recommendations should be carefully evaluated for health and safety prior to implementation.

Automation

The most ideal solution to ergonomic problems in which workers are required to manually handle materials or perform repetitive tasks is usually to automate the processes. When possible, let the workers perform a control or monitor function and let a machine do the physical work. Automation can range from single conveyors to very specialized robotics. Many companies can easily justify the cost of a conveyor or lift tables, but very few even consider using robots solely because of the initial investment cost. Automation of any kind does require additional training for workers and careful steps to ensure that the automation introduced will not itself create a safety or health hazard.

Engineering Redesign

Other redesign options include work station, task, and tool redesign. Work station redesign may involve adjusting the work height or orientation and can often enable workers to keep all reaches within a comfortable reach envelope (16 in to 18 in) and within plus or minus 2 in of elbow height. Often industrial chairs are inadequate, so careful chair selection and adjustability for worker comfort, stability, and lumbar support is important. Manual lifting and handling aids are often a very valuable part of work station redesign. Dollies, roller bins, frictionless conveyors, and hydraulic lift tables can be used to make jobs easier when automation is not practical or feasible.

Task redesign involves changing the methods of performing a job to reduce the force, repetitiveness, and/or use of awkward postures. For example; redesigning an item so that instead of workers screwing parts together, they can snap them together. Lengthening a task cycle, reducing line speed, and introducing buffers are other ways to reduce the number of repetitions involved in task performance. Task redesign may also involve changing the order in which job components are completed, dividing a task between two workers, or changing production specifications.

Tool Redesign

Tool redesign is needed when the handle of a tool ends in the palm of the hand and when the orientation of the work causes workers to use awkward postures. In many cases, tool redesign is actually tool reselection since ergonomically sound tools are commercially available for a wide variety of applications. Pistol grip and straight grip tools are available for many jobs, and should be selected to allow workers to accomplish the task with a straight wrist. Automatic clippers and spring loaded scissors can often be used to reduce the amount of stress involved in manual cutting tasks.

Tool handles have been designed to meet a wide variety of needs, such as, different sized hands, thermal insulation, reduced vibration, palmar support, and torque reduction. Generally, tool handles should be at least 4 in long to accommodate the majority of the working population, and for cylindrical handles the optimum handle diameter for force output is 1.5 in.

Counterbalancing systems are available to offset the weight of the tools themselves. As a rule of thumb, this becomes important where the weight of the tool exceeds 4 lb. A feasible, ergonomically sound tool selection alternative is usually available for inadequate or inappropriate tool use. Highly specialized tools, which are not readily available, must be developed in-house or special-ordered from an outside company. The cost can be very high and often prohibitive.

Redesign efforts usually meet with some initial resistance from affected workers. One way to reduce resistance is to involve workers in the redesign process from the beginning, by having them provide input about their needs and concerns. Worker training and feedback is vital whenever workplace changes are made. Over time, workers will almost always accept changes that make their jobs easier, especially if they are involved in the design process. Again, the best way to implement changes is with worker input and feedback.

ADMINISTRATIVE CONSIDERATIONS

In addition to redesign intervention, several administrative activities may help reduce the incidence of musculoskeletal disorders, the exposure to potentially harmful tasks, and the severity of previously discovered cumulative trauma disorders. If the source of the ergonomic problem is not eliminated or mitigated, however, the problem will likely continue or reappear. Consequently, administrative controls should be an adjunct to, not a replacement for, engineering control.

Training

Training is often insufficient and in some cases non-existent. Training should begin with safe, sound job methods. Through task analysis, companies should identify the safest ways to perform tasks in their plants, and workers should be taught the safest way to do the job from the beginning of their employment.

Adjustability and other tool or chair features should be reviewed with workers before they begin work. Workers involved in repetitive work should also be taught which postures contribute to the development of cumulative trauma disorders. All workers should be given formal training in basic lifting guidelines and the use of newly introduced hand tools.

Describing the anatomy of the wrist and/or back helps workers understand why there is a risk of injury. Cumulative trauma disorders have early warning signs of which workers should be aware. Hand, arm, or shoulder pains often provide enough warning to prevent serious problems from developing. Companies are encouraged to inform their workers of these symptoms. Management representatives and engineers should receive training in ergonomics as well, especially engineers who are involved in process flow and equipment design.

Medical Surveillance

Early detection programs, coupled with worker training, can help prevent the occurrence of serious disabling illnesses. Many companies consider pre-employment screening a sufficient early detection program. While pre-employment screening can determine a potential worker's history and physical condition prior to beginning work, it is not acceptable to assume that the workers' condition will not change as a result of the work.

When workers complain of wrist or joint pain, the complaint should be taken seriously and measures should be taken to prevent the condition from worsening. Initial complaints of wrist pain tend to be completely ignored. As a result, those complaints often escalate into serious problems over a relatively short period of time. When this happens, a change in task requirements, either through job scheduling or light duty, is recommended. A doctor's advice and care should be solicited for handling cumulative trauma cases. Remember, if the source of the problem is not designed out of the precess, the probability of illness recurrences remains high should the worker return to the same job and exposure.

Light duty programs are essential for effective handling of musculoskeletal injuries, particularly back injuries. Light duty programs require all workers to perform tasks that do not have the same demands as their regular jobs while they recover from injuries or illnesses. Light duty tasks may provide a way to keep workers on the job or get them back to work safely after an injury or illness. A doctor's approval of light duty tasks for injured or returning workers is crucial.

Work Scheduling

Work scheduling is a valuable ergonomics tool. Job rotation, break schedules, and shift length can be designed to help reduce the risk of CTDs. Job rotation must consider the risk factors associated with many different jobs to minimize the amount of time spent at potentially harmful jobs. Companies should carefully evaluate each job on a rotation schedule, and design the schedule so that workers will not rotate from one highly repetitive task to another very similar task. As a rule, tasks on the rotation schedule should have different strength requirements and should require the use of different muscle groups and limbs.

The task requirements should be considered when determining the lengths of shifts. Although extended workdays are a common practice, recovery periods allowed are usually designed for shorter eight-hour work shifts. Breaks during the workday should be based on shift length and the type of work being done, not solely on standard practice.

Standard breaks, 10 to 15 minutes in the morning and afternoon and 30 minutes for lunch, may not be adequate for extended workdays and highly repetitive work. Very few companies provide for breaks in addition to the standard break schedule. Extra relief time should be designed into repetitive tasks or extended work schedules.

A combination of redesign and administrative measures can reduce the frequency and severity of musculoskeletal injuries over time. Ergonomic workplace improvements should be implemented with support and input from top management, personnel, safety, and health professionals, engineers, maintenance staff, and affected workers. Further, modifications should be implemented on a trial basis and evaluated for effectiveness before full scale changes are made permanently.

WHERE DO WE GO FROM HERE

In several cases a decrease in the incidence of back injuries has been achieved through the installation of simple flip tables, lift tables, and extended conveyors. One company reported that the incidence of carpal tunnel syndrome cases has dramatically declined after introducing some very simple automation. Personnel and safety managers at many companies report that knowing the basic principles of biomechanics and ergonomic design has enabled them to see the need for ergonomic improvements throughout their plants. Most feel that ergonomics applications will be vital for the continued success and future of their companies.

A significant number of companies which are now making or have recently made improvements, are finding that modifications have come too late. The number of new employee occupational illnesses seems to be decreasing, but there are still long-time workers who are suffering from the effects of years spent in poorly designed workplaces. This demonstrates the need to respond to ergonomic problems in a proactive, preventive manner whenever possible instead of reacting to a problem that has already occurred.

As companies continue to become more aware of the ergonomic needs in their workplaces, the number of those requesting ergonomic assistance is growing at a very rapid pace. Emerging ergonomic issues include the impact of advanced manufacturing technology on human health and performance and workplace modifications for an aging workforce and a growing number of women in the workforce.

With increasing education, collaboration, and cooperation, companies will begin to apply ergonomic principles at the conception of processes, where they will have greatest impact on the prevention of occupational injuries and illnesses. While the areas of primary concern and the challenges will change dramatically over the next decade, companies' desire to provide a safe and healthy workplace for workers will only increase, and ergonomic applications will become an increasingly necessary part of every company's safety and health program.

References

1. IES (Illuminating Engineering Society of North America) Industrial Lighting Committee. 1983. Proposed American National Standard Practice for Industrial Lighting. *Lighting Design and Application*. 1983;13(7): 29-68.

2. Center for Ergonomics. *Two Dimensional Static Strength Prediction Program*. Ann Arbor, MI: University of Michigan; 1986.

3. National Institute for Occupational Safety and Health. *Work Practices Guide for Manual Lifting*. Washington, DC: US government Printing Office; 1982.

4. Snook SH. The design of manual handling tasks. *Ergonomics*. 1978;21:963-985.

Daniel J. Ortiz is Associate Head of the Environment Health, and Safety Division of the Georgia Institute of Technology.

Susan M. Gleaves is a Consulting Engineer with Auburn Engineers, Inc. She has both a Bachelor and Master of Science in Industrial Engineering from Auburn Univresity. Since joining Auburn Engineers, Susan has been a consultant for OSHA and has conducted ergonomics training for OSHA compliance officers, as well as providing ergonomic and safety consultations to a wide range of industrial clients. Susan's current interests are concentrated on occupational ergonomics, with a special emphasis on cumulative trauma disorders, workplace design, worker/equipment compatibility, and ergonomics/safety interactions.

18

Packing Operation in the Put-up Department: An Ergonomic Study

Carl R. Lindenmeyer
Robert S. Willoughby

This study of labor-intensive work at a textile plant was prompted by worker complaints. Packers of printed fabric rolls complained of possible long-term physical effects caused by their jobs; they also disagreed with fabric roll winders at the plant over the total yardage per roll. The winders wanted more yards per roll so they could take full advantage of the company's production incentive plan. Although they shared in the incentive plan, the packers wanted less yards per roll for ease in handling. Plant management wanted to resolve this problem without sacrificing productivity.

To start, we viewed three video tapes supplied by the company as well as making a fourth of the packers' tasks. The job evidently involved inefficient double handling and highly repetitious movement. The weight the packers were required to lift and carry was apparently excessive.

PERSPECTIVES: MANAGEMENT-WORKERS

We interviewed the plant management and engineering and discussed the problem in detail. The packers indicated that handling and packing wide rolls (up to 60 in) that contained in excess of 100 yd of fabric caused physical distress. Currently, the limitation of yardage on a wide roll is left up to the winders and the packers to determine. This approach has not been successful, resulting in conflict between the winders and the packers on what should be the maximum yardage per wide roll.

Interviews with union leadership and each of the five packers revealed a number of specific problems in the packing operation, including:

- Various physical problems, such as distress in the arms, lower back, wrists, shoulder/trapezius muscles, and legs. In addition, one packer complained of finger tingling and occasional pain when spreading his fingers.
- Easels (the main material handling device) were not always in good repair; specifically, the wheels were not rolling easily.
- One packer suggested that the packing area layout, carton area, and set-up operation could be improved.
- Other problems included the need to improve illumination, provide a stool in the packing area for use during idle periods, and adjust the height of the tables in the packing area.

All of the workers interviewed were cooperative and made valuable comments regarding the problem. The consensus among the packers was that limiting the wide rolls to 100 yd would solve their problem. A number of possible solutions were discussed with the packers as alternatives to the yardage limitation, including an adjustable height easel.

Each packer was cautioned not to twist the spine when moving rolls to the easel or the carton, but to turn the entire body. Additionally, the packers were advised to keep a straight wrist when possible and to avoid the two bad wrist angles—ulnar deviation and palmar flexion.

IN-PLANT CONSULTATION WRAP-UP

An exit meeting with management and union leadership addressed the above problems and outlined the solution strategies to be studied and reported in detail in a final report. These solutions and problem-solving approaches included:

- Shading (which ensures proper color match of fabric rolls) and packing of fabric rolls directly off the winding tables to avoid double handling
- Redesigning the easel to provide adjustable height
- Providing training in safe material handling along with roll yardage and stack height limitations
- Conducting a full Job Severity Index[1, 2] or a NIOSH formula study[3]

PRESENT METHOD

From the three videotapes supplied by the company and the fourth videotape made during the initial plant visit, we charted the current packing operation task sequence, distances, and times. We also noted opportunities for improving material handling and reducing packer stress (Figure 18-1). The present layout, keyed to the flow process chart, is shown in Figure 18-2.

FLOW PROCESS CHART	1. NUMBER	2. PAGE NO.	3. NO. OF PAGES
	1	1	1

4. PROCESS	PACKING OPERATION
6. [X] MAN [ ] MATERIAL	
7. CHART BEGINS	AT PACKING STATION
8. CHART ENDS	AT PACKING STATION
9. CHARTED BY	R.S. WILLOUGHBY
10. DATE	1/8/88
11. ORGANIZATION	PRINT WORKS

5. SUMMARY

a. ACTIONS	b. PRESENT NO.	b. PRESENT TIME	c. PROPOSED NO.	c. PROPOSED TIME	d. DIFFERENCE NO.	d. DIFFERENCE TIME
○ OPERATIONS	6					
⇨ TRANSPORTATIONS	6					
□ INSPECTIONS	2					
D DELAYS	0	--				
▽ STORAGES	0	--				
DISTANCE TRAV. (FEET)	66'					

12a. DETAILS OF [X] PRESENT [ ] PROPOSED METHOD

12a.	b. OPERATION / TRANSPORTATION / INSPECTION / DELAY / STORAGE	c. DISTANCE IN FEET	d. QUANTITY	e. TIME	f. ANALYSIS WHY? WHAT?	WHERE?	WHEN?	WHO?	HOW?	g. NOTES	h. ANALYSIS ELIMINATE	COMBINE	CH SEQUENCE	CH PLACE	CH PERSON	IMPROVE
1. PUSH EASEL TO SHADE TABLE	○ ⇨ □ D ▽	25'	1	.07						DISTANCE VARIES BETWEEN 15' & 35'	X					X
2. SELECT ROLL (SHADE)	○ ⇨ □ D ▽	-	4.4	↓						ROLLS MATCHED FOR COLOR (TIME COMBINED WITH #3)						
3. PUT ROLL ON EASEL	○ ⇨ □ D ▽	-	4.4	.65						PICK UP ROLL, PUT ROLL ON EASEL (AVG. 4.4 ROLLS)	X					X
4. PUSH EASEL TO PACKING STATION	○ ⇨ □ D ▽	25'	1	.12						DISTANCE VARIES BETWEEN 15' & 35'	X					X
5. WALK TO CONVEYOR (CARTON STORAGE)	○ ⇨ □ D ▽	5'	1	↓						CARTONS STORED ABOVE CONVEYOR (TIME COMBINED WITH ELEMENTS 6 & 7						
6. ASSEMBLE CARTON	○ ⇨ □ D ▽	-	1	↓												
7. WALK TO EASEL	○ ⇨ □ D ▽	5'	1	.72												
8. SELECT ROLL	○ ⇨ □ D ▽	-	4.4	↓						TIME FOR ELEMENTS 8, 9, 10, 11, AND 12 COMBINED						
9. PUT ROLL IN CARTON	○ ⇨ □ D ▽	-	4.4	↓						PICK UP ROLL, PUT IN CARTON (AVG. 4.4 ROLLS PER CARTON)	X					X
10. WALK TO CALCULATOR	○ ⇨ □ D ▽	3'	4.4	↓							X					X
11. RECORD YARDAGE	○ ⇨ □ D ▽	-	4.4	↓												
12. WALK TO EASEL	○ ⇨ □ D ▽	3'	4.4	.75							X					X
13. COMPLETE PAPERWORK	○ ⇨ □ D ▽	-	1	.40						TOTAL YARDAGE, LABEL CARTON						
14. MISC. PRODUCTIVE WORK	○ ⇨ □ D ▽	-	1	.35												
15.	○ ⇨ □ D ▽															

[Margin brackets labelled "combined": elements 2–3; elements 5–7; elements 8–12]

Figure 18-1. Packing Flow Process Chart.

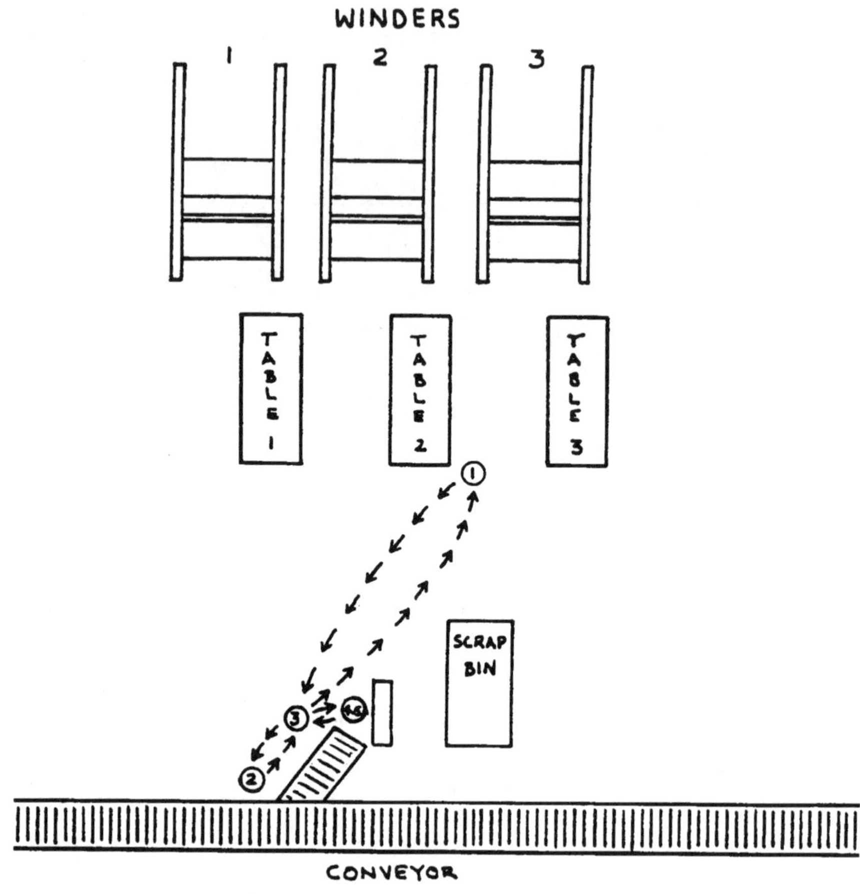

Figure 18-2. Packing Flow Diagram.

These charts graphically portray the human stress problems of excessive lifting and double handling, and reveal excessive movement involved in paperwork. They are important aids in understanding the problems surrounding the packing operation and in generating the solutions to these problems.

WORK ASSESSMENT

Even though we had qualitatively determined that the job needed significant ergonomic improvement, we decided to provide the company with a quantitative assessment of the packing operation based on biomechanical, physiological, and psychophysical (lifting limits based on worker self-assessment) research. This approach is simply evaluated from the current manual material handling workloads. We considered two approaches to performing this analysis.

Job Severity Index

Liles et al[2] have developed a formula that calculates a Job Severity Index (JSI) for any given lifting task. This formula takes into account certain task variables, such as weight, frequency of lift, container size, and range of lift, along with certain job exposure factors, such as length of the work week and work shift. Individual worker characteristics, such as body weight, arm strength, back strength, and age are also considered. The results of these authors' research included the JSI formula and a JSI threshold value (1.5), above which the incidence, severity, and cost of injury dramatically increases.

This method of job assessment requires collecting a lot of data. Individual worker capacities as well as job characteristics need to be measured. Such a JSI assessment study could prove time consuming as well as costly. For a group of workers, however, the technique can be used with relative ease.

NIOSH Formula

The National Institute of Occupational Safety and Health (NIOSH) has developed a guideline formula for estimating the risk of various combinations of task variables.[3] Underlying this formula are the concepts of *maximum permissible limit* (MPL), the maximum limit that should be allowed, and *action limit* (AL), the limit above which some administrative and/or engineering control of the task is required.

The formula estimates the action limit. The MPL is simply three times the AL. Because of past experience with this formula, we decided to use the NIOSH formula instead of the JSI. Before we applied the NIOSH formula, however, we outlined the job elements that could produce injury.

JOB DESCRIPTION

Each fabric packer services three winders. The packers lift rolls from the winding tables, move them to an easel, and lift or toss the rolls onto the easel. After obtaining a number of rolls, usually from 5 to 12, the packers push the easel to their station at the main conveyor. The packers assemble a carton by pulling two halves of a cardboard carton from storage racks located above the conveyor, opening the halves up, and sliding each half onto the other. Then, they lift the rolls from the easel and drop or toss them into the carton. The entire cycle involves lifting each roll twice (double handling). Figures 18-3 and 18-4 illustrate lifting to the easel and the roll handling to the carton portions of the job.

Lifting Rolls from Winding Table to Easel

The first part of the job requires the packers to lift a fabric roll from the winding table to an easel. The initial lift starts at a vertical height (V_i) of about 35 in above the floor with the hands about 12 in from the packers' center of gravity (H_i). At the end of the lift, the typical vertical height of the hands is approximately 47 in (V_f), and the distance between the packers' center of gravity and the load increase to about 18 in to 24 in (Hf). Figure 18-5 shows the packers' movement and the various distances necessary to use the NIOSH formula.

Figure 18-3. Lifting Rolls from Winding Table to Easel.

Lift Rolls from Easel to Carton

After the stacked easel is pushed to the packing area, the fabric rolls are packed into cartons. The average number of rolls packed per carton is 4.4. The dimensions as described for the previous task are: V_i = 38 in, H_i = 12 in, V_f = 50 in, and H_f = 24 in, as illustrated in Figure 18-6.

Frequency, Weight, Size of Lift

The average number of rolls produced per winder is estimated at 31 per hour, and since each packer services three winders, a packer, on the average, handles 93 rolls per hour. The weight per roll ranges from about 2 lb to 70 lb with an average weight of 47 lb. The length of rolls range from 45 in to 60 in, with 65% of the rolls measuring 45 in in length.

Other Potentially Injurious Motions

The packers consistently turn their upper body while lifting the rolls. This twisting motion produces torsional stresses in the spine and should be avoided

Figure 18-4. Lifting Rolls from Easel to Carton.

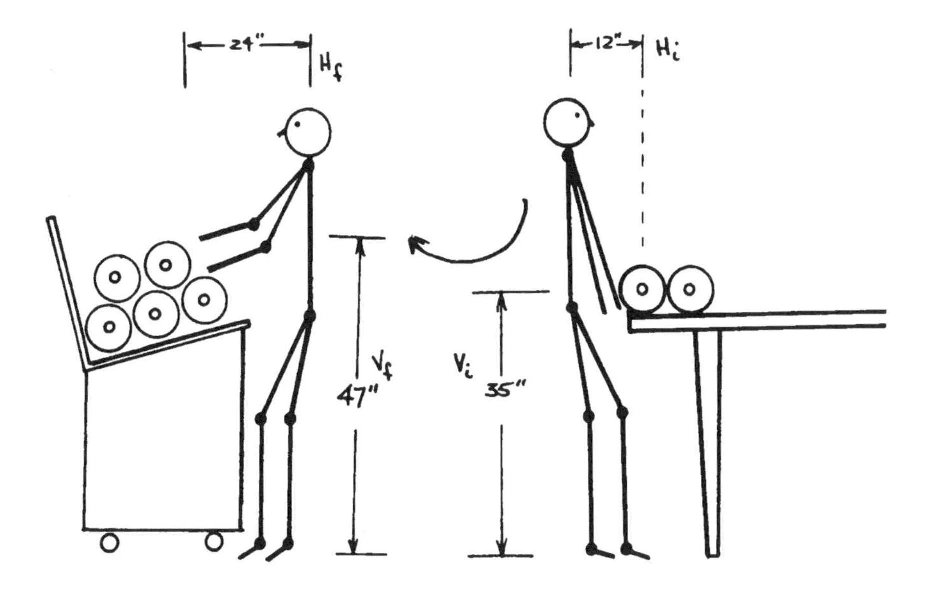

Figure 18-5. Lifting Rolls from Winding Table to Easel (NIOSH Data)

through engineering design and/or administrative controls, such as training and/or weight limitations.

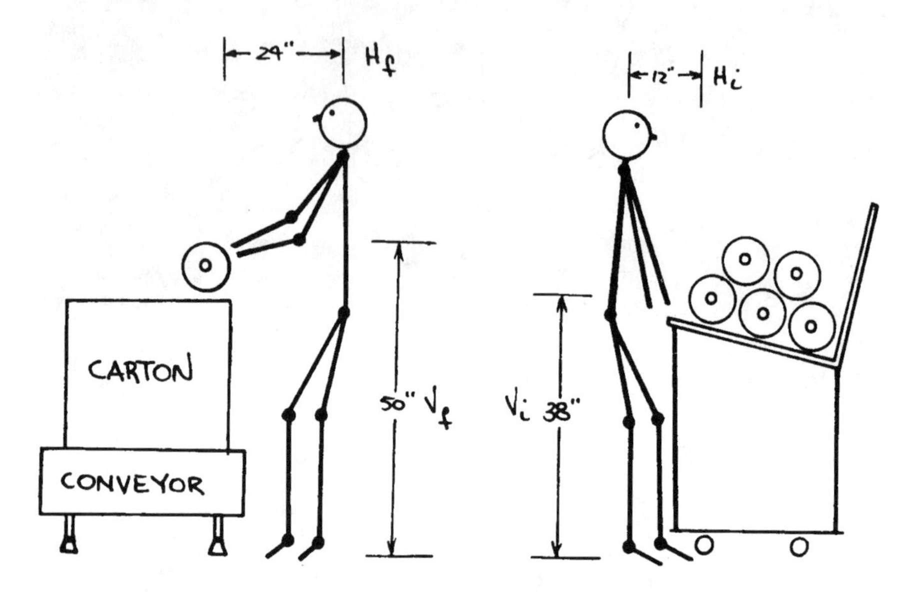

Figure 18-6. Lifting Rolls from Easel to Carton (NIOSH Data).

We observed that the packers do not lift loads symmetrically, especially when handling the wide rolls. Lifting loads asymmetrically causes the spine to bend to one side, resulting in a lateral bending movement on the lumbar column. Lifting loads asymmetrically also produces a rotation of each vertebra on its adjacent vertebra (similar to twisting).

When packers pick-up rolls from the winding table and place them on the easel, they frequently bend their wrists. Frequent wrist deviations, particularly ulnar deviation and/or palmar flexion, can result in cumulative trauma disorders. The methods observed could result in tenosynovitis, which is inflammation of the tendons and tendon sheaths. Carpal tunnel syndrome can also develop, which results in damage to the median nerve.

When the packers start to lift a roll, especially a heavy or over-sized roll, the force exerted should be applied in a smooth increasing fashion. The dynamic forces imparted by the observed rapid or jerky motions can greatly increase the load's negative effect.

When putting together the packing cartons, the packers' arms extend 12 in to 25 in from their bodies while exerting a slight force to pull the carton out of the rack. In addition, their arms are about shoulder height. The packers maintain this position while assembling the cartons.

These motions are not particularly injurious, but they may cause excessive fatigue over what would be encountered if the packers worked with their hands closer to their bodies and at about waist level. When these stresses are added to the other significant job-related stresses, fatigue increases. The body's ability to handle stress before incurring injury decreases as fatigue sets in, particularly due to stresses on the spinal column.[3]

APPLYING THE NIOSH FORMULA

The lifting limits, estimated by the NIOSH formula, are intended to apply where there is smooth, two-handed symmetric lifting in the sagittal plane (object moves parallel to the vertical body line), moderate object width (30 in or less), unrestricted lifting posture, good couplings (handles, shoes, floor surface), and favorable ambient environments. Two weight limits are calculated, the action limit, and the maximum permissible limit. Jobs with lifting limits of AL or less can be performed by almost any male worker and most female workers. Jobs with lifting limits of MPL can be done by only a few people.

NIOSH recommends that administrative and engineering controls be imposed on jobs with lifting limits greater than the AL, and recommends that jobs with lifting limits greater than the MPL be redesigned. The NIOSH formula contains no safety factors, even though engineering formulas often contain such factors.

The Formula and Task Variables

The NIOSH formula is shown below:

AL (lb) = 90x(6/H)x(1-(.01xABS(V-30)))x(0.7+3/D)x(1-F/Fmax)
MPL(lb) = 3xAL
where:
AL = action limit
MPL = maximum possible limit
H = horizontal location (inches) forward of midpoint between ankles at origin of lift
V = vertical location (inches) at origin of lift
ABS = absolute value
D = vertical travel distance (inches) between origin and destination of lift
F = frequency of lift (lifts/minute)
Fmax = maximum frequency which can be sustained
= 18 if occasional high frequency and standing
= 15 if continuous high frequency and standing
= 15 if occasional high frequency and stooping
= 12 if continuous high frequency and stooping

The two types of lifting tasks defined for the packing operation are similar enough to combine together. So, recalling the statistics that describe the packing operation in terms of lifting, the following shows how the NIOSH formula can be used to estimate the action limit and the maximum permissible limit for this job:

H = 12 inches V = 36 inches
D = 12 inches F = 3.1 lifts/minute
Fmax = 15 lifts/minute
AL (lb) = 90x(6/12)x(1-(.01xABS(36-30)))x(0.7+3/12)x(1-3.1/15)
MPL (lb) = 3xAL
AL = 31.9 lbs
MPL = 95.7 lbs

As noted, the actual range of weight lifted is from 2 lb to 70 lb, with an average weight of 47 lb. Also, many of the limitations of the NIOSH formula are being violated in this job; for example, no twisting, smooth lifting, good couplings/handles, moderate object width, and symmetric lifting. Considering these violations, the actual AL and MPL are probably at best moderately lower to at worst significantly lower than calculated. The actual 70 lb value is significantly above the AL (31.9 lb), indicating administrative and/or engineering controls. Although the actual 70 lb value is below the MPL (95.7), twisting, asymmetric lifting, and the other formula deviations makes this job a strong candidate for engineering redesign.

ADMINISTRATIVE CONTROLS

One possible solution includes training and a yardage limitation imposed on wide rolls. This solution would address the ergonomic problem through education in proper manual material handling methods and motions, and would reduce worker stress through limiting the weight of rolls handled. Roll size and weight limitations may unnecessarily restrict productivity and worker earnings, however.

Training

Packers should be trained in proper lifting techniques to avoid back and other injuries. There apparently is not a significant amount of back bending or stooping with heavy weights involved in the packing operation. Some consideration, however, should be given to the body turn motions during the loading of the easel and during the carton packing. In these two tasks, the packers could possibly twist the spine, rather than perform a full turn involving the entire body.

During the observations of the video tape, we noted that only some of the packers normally used a Turn Body Case 1 (TBC1 in MTM); that is the lagging foot was not brought up adjacent to the leading foot. On a few occasions, we noted that a full Turn Body Case 2 (lagging foot brought up) was used. Either of these full body turns are recommended to avoid twisting of the spinal column. Twisting the spine repeatedly can cause degeneration of the vertebral discs that act, in part, as cushions between the vertebrae.

In addition, we recommended that the packers be trained and periodically advised to maintain a straight wrist during the entire packing cycle. Training the packers to keep their wrists straight may be difficult, because they can easily forget to try to do this and the job itself requires unnatural wrist angles. We also suggested that video tape recordings be made of incorrect as well as proper methods of roll handling, lifting, and carrying. These video tapes can then be used for training in safe material handling practices, at least until engineering redesign changes are implemented.

Yardage Limitations

Obviously, a yardage limitation on wide rolls would reduce worker stress and the risk of injury. The packers indicated that a company-imposed 100 yd limitation would satisfy their complaint. The winders would then be dissatisfied, due to limitations on their earnings. Further, if the company's customers normally accept larger yardage rolls, and rolls are sent with limits of 100 yd, the customer will experience higher operation costs. Consequently, more rolls will be sent to satisfy the order, resulting in more roll-end scrap and more roll handling, resulting in higher material handling costs.

Although this is not an optimal solution, we recommended that a 100-yd limitation be imposed on wide rolls as a temporary measure. Further, a maximum easel stack height of three or at most four rolls is also recommended as a temporary measure until a new material handling system is implemented.

ENGINEERING DESIGN SOLUTIONS

Solution A

This solution involves the parallel relocation of the main packing line conveyor to a position closer to the winding tables. The cartons on a ball roller bed transfer cart would be packed directly off the winding tables. Lifting is virtually eliminated since the rolls are rolled off the winding table directly into the carton. Double handling is also eliminated. See Figure 18-7 for the overall layout of Solution A and Figure 18-8 for a detailed drawing of the ball roller bed transfer cart.

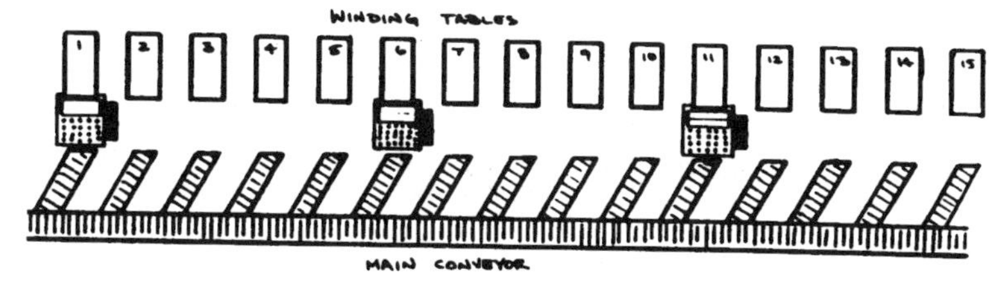

Figure 18-7. Overall Layout for Design Solution A.

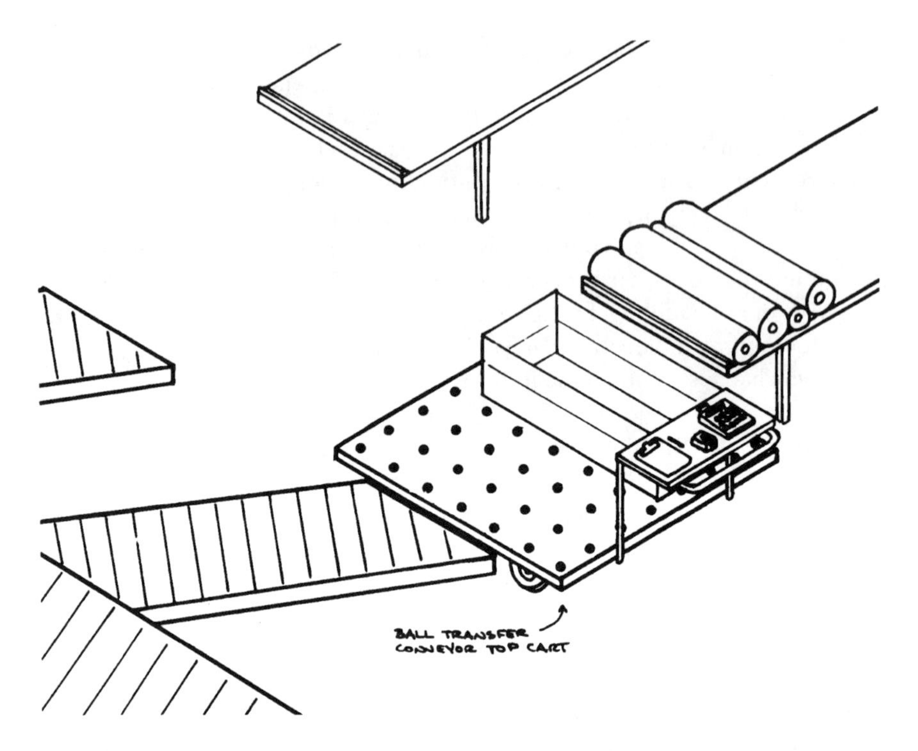

Figure 18-8. Detail Drawing of Transfer Cart for Design Solutions A and B.

Solution B

This solution is identical to Solution A, with the exception that the main packing conveyor is not moved from its present location. The transfer cart shown in Figure 18-8 would be used. Longer conveyor spurs are provided as illustrated in Figure 18-9. This solution has the same ergonomic (lifting virtually eliminated and no double handling) advantages as Solution A. However, the packers would have to push the cartons along the longer spur conveyor to the main conveyor, resulting in more time and effort.

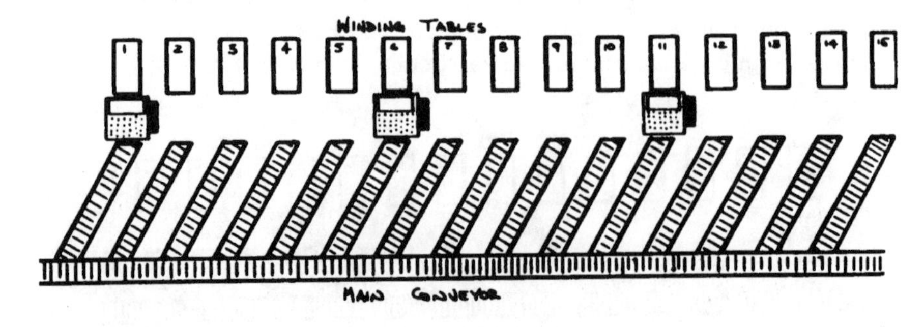

Figure 18-9. Overall Layout od Design Solution B.

Solution C

This solution involves the use of a ball roller bed cart, similar to the one described in Solutions A and B. With this solution, the rolls are shaded and packed directly off of the winding table to a carton placed on the ball roller bed of the cart. The carton is filled, taped, and paperwork done at the winding table. The cart is then pushed or pulled to the main conveyor where the carton is pushed onto the spur and then onto the main conveyor line. The overall layout for this solution is shown as Figure 18-10. A detailed drawing of the cart interfacing the winding table is shown as Figure 18-11. Figure 18-12 shows the cart in position at the main conveyor spur.

As with the previous two design solutions, this solution virtually eliminates lifting of the rolls, by having the packers roll and drop them directly into the cartons at the winding table. Also, double handling is eliminated, although one round trip of the cart is required. Work effort reduction and ergonomic improvement, therefore, is not as great as in Solutions A and B.

The recommended cart can be designed to carry up to three cartons, which will reduce packer effort by minimizing the number of round trips to the main conveyor per carton handled. This solution has two distinct advantages over Solutions A and B. First, there will be no lost production time for installation. Second, worker and other material handling traffic through the area would continue to be permitted.

Solution D

This solution was discussed with the packers to get their ideas and reactions. The easel would be designed with an adjustable height feature to permit rolling or dropping of the rolls from the winding table to the easel. The lower height would permit at least three levels of rolls without lifting. The filled easel in down position would be pushed to the main conveyor spur. The height would be raised by a pneumatic lift device to facilitate rolling or dropping the rolls into the carton in a similar fashion to that which is done at the present time. A detail drawing of the redesigned easel incorporating a scissors-type pneumatic lift device is shown as Figure 18-13.

The solution eliminates the worker stress from excessive lifting and twisting of the spine. With proper easel design and appropriate overhand of the horizontal easel surface, the rolls could possibly roll or drop directly from the easel to the carton. More, detailed design work would be required if this solution was pursued.

An Additional Design Idea

Since solutions A, B, C, and D require the packer to roll or drop the fabric roll from the winding table to the open carton on the ball roller bed transfer cart or to the redesigned easel, an additional device was designed. By using a foot operated, spring-loaded hinged, *drop leaf* extension to the winding table, it is possible to eliminate this effort (Figure 18-14). A foot motion would be required

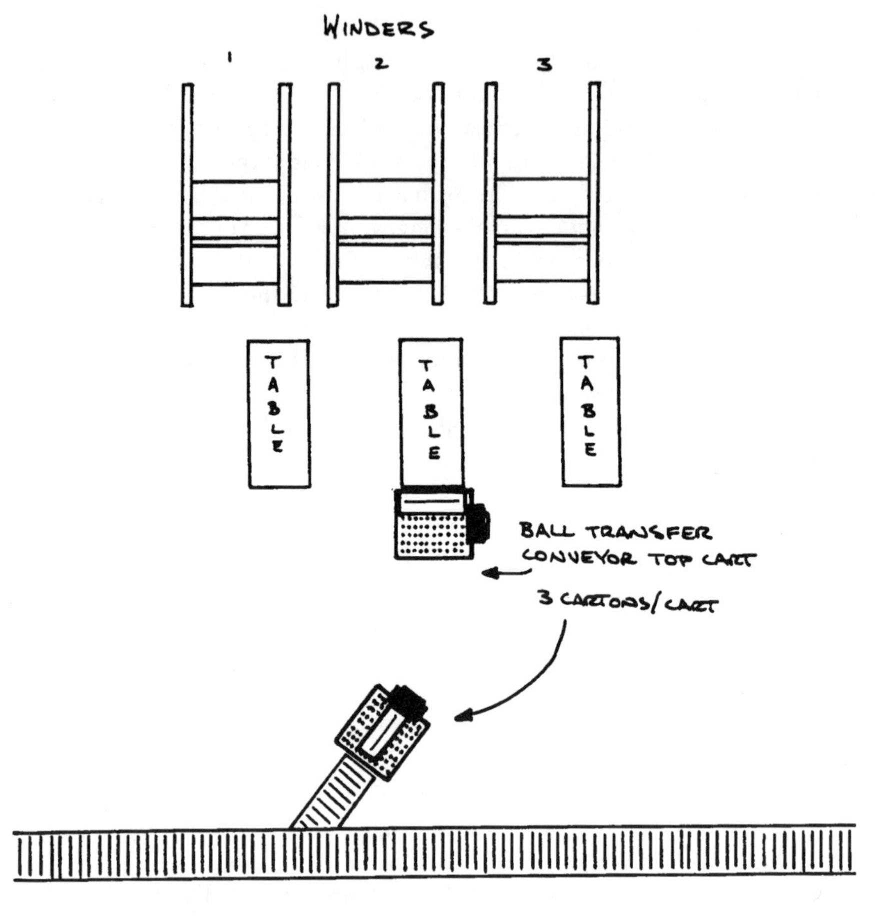

Figure 18-10. Overall Layout of Design Solution C.

to press the foot pedal. This additional improvement would positively impact the ergonomics of the packing operation. For Solutions A, B, and C, it reduces stress in packing cartons at the winding table. For Solution D, it reduces stress in loading the easel at the winding table. Further, if the drop leaf concept were to be applied to the easel, carton packing at the main conveyor would be significantly less stressful.

Advantages of a Design Solution

Ergonomic problems can be solved a number of ways including:

- Selecting workers who can do the task without undue strain. This is not normally possible due to equal opportunity legislation and the need to develop accurate and defensible methods to determine job requirements, including tests and standards for the selected workers to satisfy.

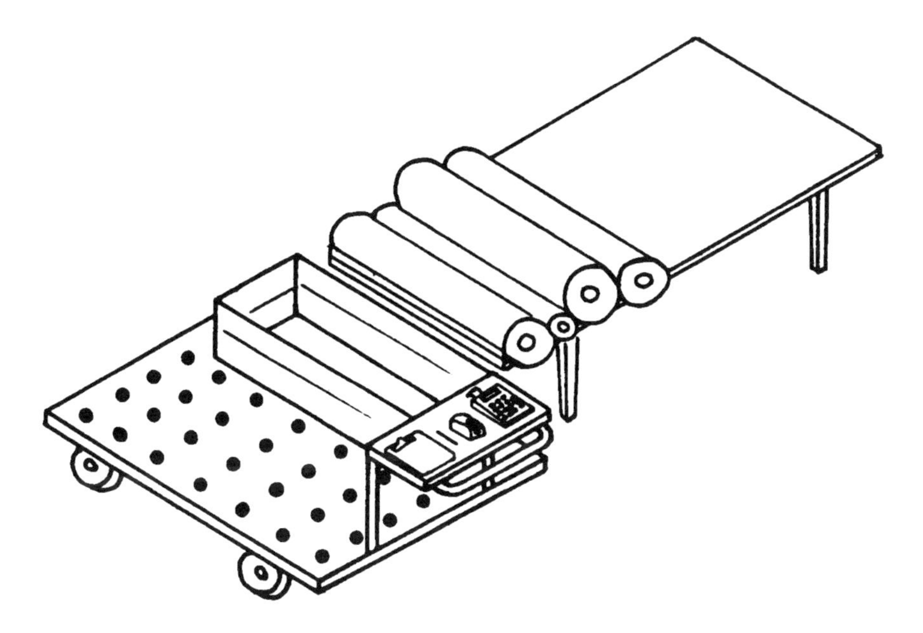

Figure 18-11. Detail Drawing of Cart at Winding Table for Design Solution C.

- Training workers in the correct method to perform the job. Proper methods in manual material handling, including lifting, are demonstrated, practiced, and a corrective action program used to ensure compliance. This approach, although essential for all potentially hazardous jobs, has two limitations. One, the workers may forget (or not apply) the correct methods. Two, the job may be so demanding or hazardous that even the best methods have a significant potential for injury. This approach can also include physical training/exercise.
- Implementing other administrative controls and establishing limits, such as weights lifted and stack heights. Such limits should be considered temporary and only used until real design solutions are implemented.
- Redesigning the job to eliminate, or greatly reduce, the ergonomic problem. Designing the work method and motions, along with mechanical layout and material handling equipment design, represents an effective and permanent solution. Further, this approach allows us to "have our cake and eat it too" in that the ergonomic problem is solved, workers do not have their earnings limited, and productivity is not compromised. In fact, productivity may even be improved. To coin a new term, we practice *ergonomic economics*.

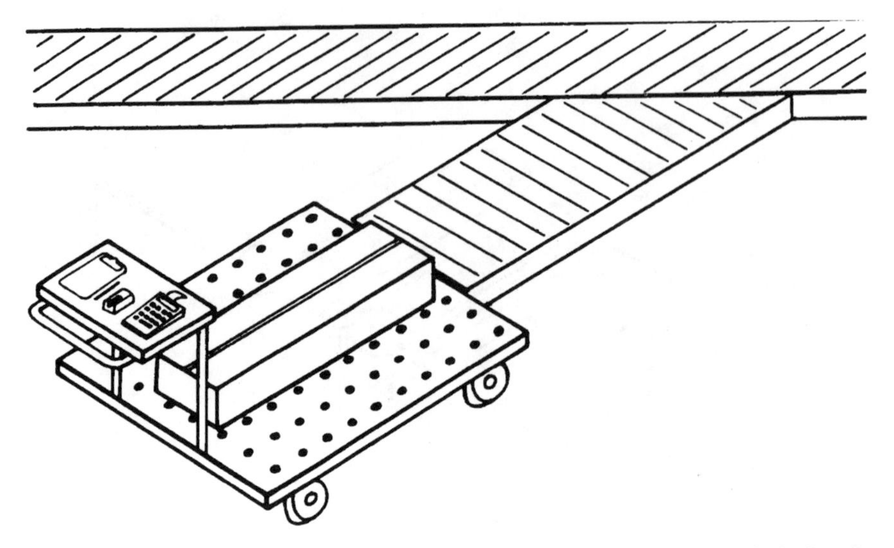

Figure 18-12. Detail Drawing of Cart at Main Conveyor for Design Solution C.

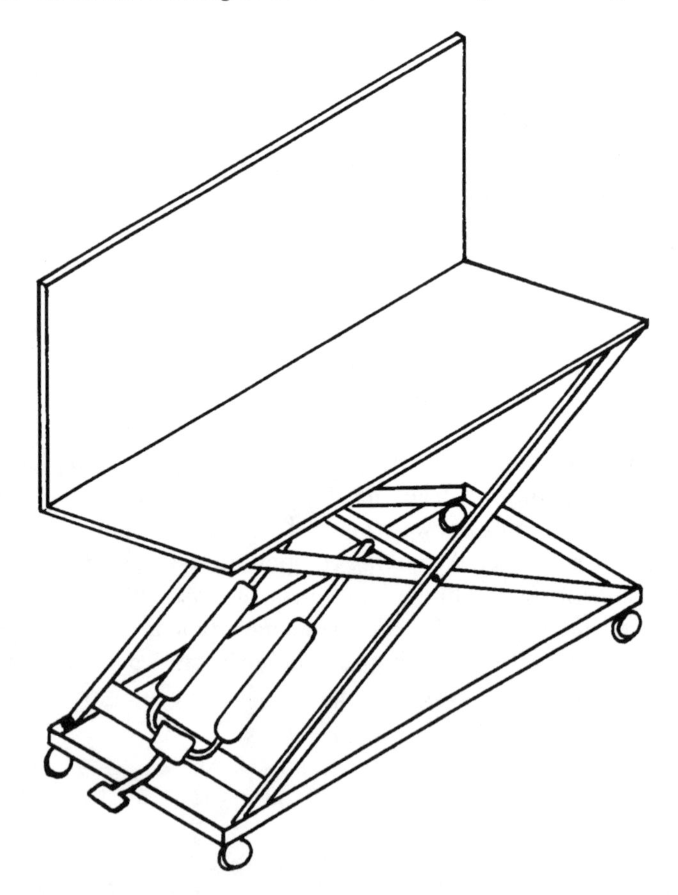

Figure 18-13. Detail Drawing of Redesigned Easel for Design Solution D.

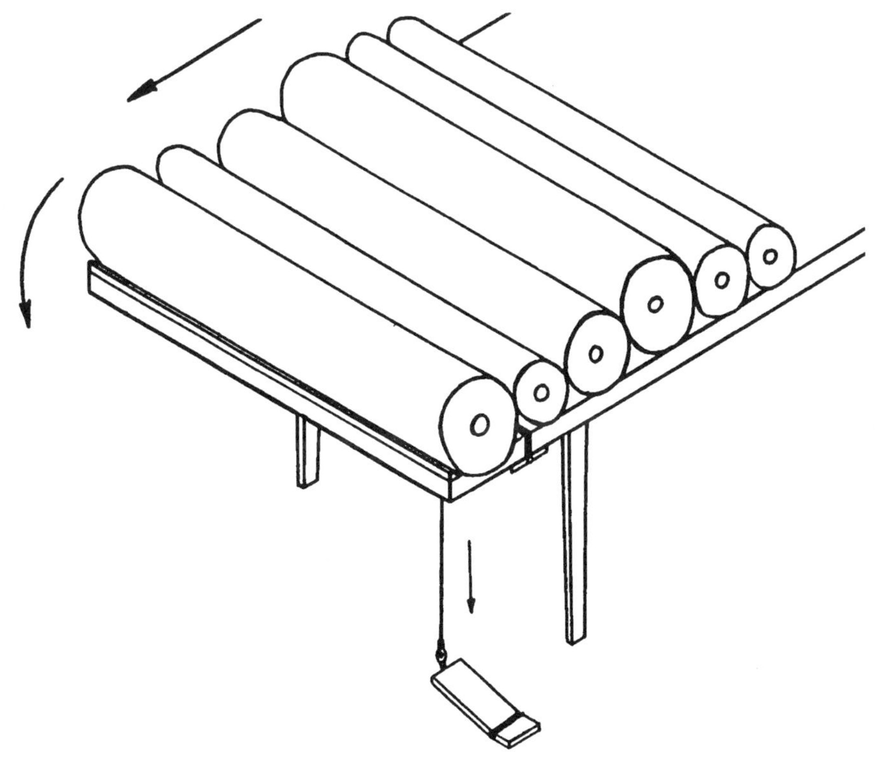

Figure 18-14. Drop Leaf Mechanism for Winding/Shading Table.
Note: When the foot pedal is stepped on, the drop leaf portion of the winding table is tilted down about 15° and the fabric roll falls directly onto the carton waiting below. Spring loaded hinges permit the drop leaf to return to the horizontal position and are strong enough to hold up to a 70 lb roll.

TIME AND COST ANALYSIS

Present Method

Another outside consultant determined a packing time standard of 2.675 normal minutes per carton packed (average of five rolls per carton). This consultant's work also determined work element times that could be eliminated and/or reduced.

Time and Cost Comparison Summary

Table 18-1 shows, by work element, estimates of time to completely process one carton of five rolls using the present method as well as for Solutions A, B, C, and D. The drop leaf idea also was included in the time and cost reductions. The elements are numbered and identified as follows:

1. Push easel to shade station (winder table)
2. Shade (and put rolls on easel)

Table 18-1. Time/Cost of Solution Alternatives.

Element	Present Method Time/Carton	Solution A Time/Carton	Solution B Time/Carton	Solution C Time/Carton	Solution D Time/Carton
1	0.065	----	----	0.032	0.065
2	0.650	----	----	----	0.066
3	0.120	----	----	0.060	0.120
4	0.340	0.340	0.340	0.340	0.340
5	0.750	0.444	0.444	0.444	0.600
6	0.400	0.318	0.318	0.318	0.400
7	0.350	0.350	0.350	0.350	0.350
8	----	0.058	0.058	----	----
9	----	----	0.108	----	----
10	----	----	----	----	0.034
Total Time Per Carton	2.675	1.510	1.618	1.544	1.975
Cartons/min	0.374	0.662	0.618	0.648	0.506
% decrease in time	----	43.6%	39.5%	42.3%	26.2%
% increase in production	----	77.0%	65.2%	73.3%	35.3%
Number of packers	5	3	3	3	4
Savings/year	----	$45,000	$45,000	$45,000	$22,500
Estimated capital investment	----	$20,000	$10,000	$4,000	$6,000 to $8,000
Payback period	----	5.3 months	2.7 months	1.1 months	3.2 to 4.3 months

3. Push easel to pack station (main conveyor spur)
4. Make-up carton
5. Pack carton
6. Paperwork
7. Miscellaneous productive time (allowance)

In addition, we added three elements, which will be used on one or more of our alternative solutions:

8. Move transfer cart to next winding table (Solutions A and B)
9. Push cartons to main conveyor (Solution B)
10. Pump up redesigned scissors easel (Solution D)

Detailed Discussion of Time Analysis

Solution A. This solution affects the present time standard in several ways. In reference to Table 18-1, pushing the easel to and from the packing station is eliminated as well as the time to load the easel (Elements 1, 2, and 3). This solution does not affect Element 4, the time necessary to make a carton. The time to pack a carton is significantly reduced, however, by packing at the winding table. Table 18-2 shows a MTM analysis (a predetermined motion-time data technique), with a packing time (Element 5) of 0.475 minutes.

Element 6, paperwork, was reduced in time by eliminating the walking back and forth between the easel and the calculator table. The present time of 0.400 min per carton was reduced by 0.082 min per carton by the elimination of walking eight paces obstructed (W8PO = 136.0 TMU = 0.082 min per carton). For solution A, Element 6 becomes 0.400 - 0.082 = 0.318 min per carton. Element 7, miscellaneous productive time, is not affected. Element 8 is added to account for moving the ball roller bed transfer cart to the next winding table (0.058 min per carton). An MTM analysis was done on these work elements as well as on all the remaining existing and new work elements, using a similar analysis shown in Table 18-2.

Solution B. The element analysis shown for Solution A applies to Solution B, with the addition of Element 9, push cartons to main conveyor (on extended spur). Due to the extended length of the main conveyor spurs, the packers will be required to push the carton an additional distance and will follow the motion pattern W6PO, TBC1, and W4P for an additional time of 0.108 min (180.6) TMUs) per carton.

Solution C. This solution still requires Element 1, push easel to shade area, except the time is reduced by one-half due to the ability of the packers to deliver two cartons in one trip (0.065/2 = 0.032 min). Element 2 is eliminated as in Solutions A and B. Element 3 is still required, but reduced by one-half for the same reason as Element 1 (0.120/2 = 0.060 min). Elements 4, 5, 6, and 7 remain the same as in Solutions A and B.

Solution D. This solution, involving a redesigned adjustable height easel, requires Element 1 as presently performed. Element 2 will be significantly reduced in time due to the new, efficient motion pattern. The MTM analysis for Element 2 is, with both hands, R24B, G5, M8B5, and M12B20/2 for a total of 0.161 min (268.5 TMUs) per carton.

Table 18-2. MTM Analysis of Solution A Element 5.

LH Description	LH	f	TMU	f	RH	RH Description
Roll/Drop Roll into Carton:						
Contact grasp Position roll Move roll in	G5 M8B5 M12B20/2		0.0 13.4 18.8		G5 M8B5 M12B20/2	Contact grasp Position roll Move roll in
Tabulate Yardage:						
    Reach to roll	 R_B		18.6 ---- 7.3 29.5 18.6	 5	(TBC1 R_B EF R2C (TBC1 R_B	Turn to calculator Limited reach Eye focus Hit 5 keys Turn to table Limited reach
Tape Box (per outside consultant study):						
Tape box			52.0 (per one roll)			
Total TMU/one roll........................			158.2 TMU/roll			
Total TMU/5 roll carton..................			791.0 TMU/carton			
Total Minutes per carton................			0.475 minutes/carton			

Elements 3 and 4 remain the same as currently practiced. Element 5, packing rolls into cartons, will be reduced in time. A conservative estimate is the time will be reduced by 20%, from 0.750 min per carton to 0.600 min per carton. If the new easels are designed to accommodate the roll/drop method at the main conveyor spur, a significant amount of the present lift, carry, and drop procedure will be eliminated. We felt that a 20% reduction will be easily obtained (0.75 - 0.150 = 0.600 min per carton) due to motion elimination. Elements 6 and 7 will remain the same as in the present method. Additional time will be required to pump-up the easel before the carton is packed. This time is shown as Element 10, and consists of four 12-inch leg motions of 14.3 TMU's each for a total of 57.2 TMU or 0.034 min per carton.

The drop leaf table. This concept was developed to further reduce packers' stress and improve productivity in the packing operation. Solutions A,

B, and C require the packer to roll or drop the fabric roll from the winding table to the open carton on the ball roller bed transfer cart. The MTM motion to accomplish this is an M12B20/2 and requires 18.8 TMU per roll (Table 18-2). By using a foot operated, spring-loaded hinged drop leaf extension to the winding table, it is possible to eliminate this motion. A 4-in foot motion (LM4) of 8.5 TMU would be required. With five rolls per carton, this idea results in 5(18.5 - 8.5)(0.0006) or an additional 0.031 min saved per carton.

The time savings in Element 5 for Solutions A, B, and C will be 18.8 - 8.5 or 10.3 TMU per roll packed. With an average of five rolls packed per carton, this results in 51.5 TMU saved per carton. Savings in minutes per carton is 0.031. The new Element 5 time for Solutions A, B, and C is 0.475 - 0.031 or 0.444 min per carton.

For Solution D, involving the redesigned easel, the savings will be more significant. The reach-to-roll motion (21.5 TMU) as well as the roll/drop box (18.8 TMU) will be eliminated. Again, a 4-in foot motion (LM4) of 8.5 TMU would be required. The time savings for Solution D (in Element 2) will be 31.8 TMU (21.5 + 18.8 - 8.5 = 31.8) per roll. With an average of five rolls packed per carton this results in 159.0 TMU saved per carton. Savings in minutes per carton is 0.095. The new Element 2 time for Solution D is 0.161 - 0.095 or 0.066 min per carton.

The drop leaf table impacts the packing operation in two significant ways. It further eliminates physical work resulting in a less stressful packing operation and it enhances productivity.

CONCLUSIONS AND RECOMMENDATIONS

All four design solutions and the drop leaf table have some merit and should be thoroughly examined by management and the union for effectiveness and acceptability. The final choice for implementation will be based upon resolution of the ergonomic problem, which all four solutions will satisfy, as well as economic factors.

We recommended Solution C because of its low cost, minimum interruption to production, short payback period, and solution of the ergonomic problem. We also recommended immediate implementation of an effective job hazard analysis program as part of the company's on-going safety effort.

The company adopted Solution C with minor modifications. Further, changes in task/personnel assignments were made.

References

1. Ayoub MM, Selan JL, and Jiang BC. *A Mini-Guide for Lifting.* Lubbock, Tex: Texas Tech University; 1983.

2. Liles DH, Desvanayagam S, Ayoub MM, and Mahajan P. A job severity index for the evaluation and control of lifting injury. *Human Factors,* 1984;26:6;683-693.

3. *Work Practices Guide for Manual Lifting.* DHHS (NIOSH) Publication No. 81-122, March, 1981, 183 pp. (Available from the Superintendent of Documents, U.S. Government Printing Office, Washington, D.C. 20402).

Carl C. Lindenmeyer is Professor of Industrial Engineering at Clemson University. He earned a BSIE from Northwestern University and a MS from Western Michigan University. He has also done graduate work at Michigan State University. Professor Lindenmeyer is certified in MTM (Methods-Time-Measurement) and 4M (Micro-Matic Methods and Measurement) from the MTM Association for Standards and Research. He also holds certifications in General Purpose Data (GPD) and University MOST (Maynard Operation Sequence Technique). He is a Senior Member of both the Institute of Industrial Engineers and the American Society of Quality Control. Professor Lindenmeyer teaches, researches, and consults in the areas of methods engineering and computerized work measurement, human factors engineering, statistical process control, plant layout and materials handling.

Robert S. Willoughby is a Quality Engineer for KEMET Electronics in Greenville, South Carolina. He holds an AAS degree in Industrial Engineering Technology from Greenville Technical College and recently earned a BSIE from Clemson University. He is also a member of both the Institute of Industrial Engineers and the American Society of Quality Control. Mr. Willoughby won the "Outstanding Senior Industrial Engineer" award while at Clemson and holds memberships in the honorary societies, Tau Beta Pi, Phi Kappa Phi, and the Golden Key.

19

Implementation of Human Factors Principles in the Design of a Manufacturing Process

George J. Burri, Jr.
Martin G. Helander, PhD

INTRODUCTION

Increasingly complex manufacturing processes are placing even greater demands on workers. Additionally, these processes are creating greater interdependencies between people and machines. A systems approach to planning, which considers human factors and ergonomics early in the manufacturing system's design, is key to developing efficient manufacturing processes while ensuring worker safety and satisfaction and quality products.

Years ago, International Business Machines (IBM) recognized this and considered the worker in the early phases of the system's design process. At IBM, for example, automated manufacturing processes, which combine humans with robot work stations, emphasize the use of human factors design criteria to allocate tasks between workers and robots, while at the same time enhance job satisfaction.

This chapter reviews the literature on human factors and ergonomics as they relate to the design process. This historic perspective is helpful to understanding the significant benefits of considering human factors principles early in the system's design phase. This chapter concludes with a case study of an IBM manufacturing facility, which illustrates that benefits can be realized even if the design criteria are relatively inflexible.

HISTORIC PERSPECTIVE

In 1951, the landmark Fitts list was developed. This list contrasted the capability and limitations of people to the capability and limitations of machines. The idea was to compare what people do best with what machines do best and thereby determine optimal task allocation between people and machines.

In 1963, Jordan[1] asserted that due to capability comparisons that led to oversimplifications, little progress had been made toward developing a workable function allocation. The psychological consequences of the comparison were usually overlooked. The result was adverse effects on job satisfaction and motivation.

For example, people may perform much better than automated devices in visual inspection tasks, but these tasks are often highly repetitive and monotonous. To achieve maximum effectiveness, processes must be designed so that people and machines complement each other. Jordan pointed out that most function allocation in industry place machines and technical solutions first and people second. This is a continual cause of problems in industry; with the limitations of machines often dictating the type of work left for the human operators.

In 1980, Gerwin and Leung[2] documented a study of a flexible manufacturing system (FMS). They noted that there is still considerable debate within the corporation about the benefits of the FMS even though it has been in use for several years. The main problems turned out to be organizational rather than technical. The mechanics of installing and operating the hardware and software received most of the planning effort, while human needs and organizational issues were addressed if and when they arose.

In 1980, Nof, Knight, and Salvendy[3] discussed a job-and-skills analysis approach to optimize robot task performance. They compared in elaborate detail the abilities and limitations of industrial robots and humans, including the social and psychological needs of the human operator. They concluded that such an approach to the design of robot operations would increase productive utilization of robots in industry. This approach, however, disregards any problems caused in implementing organizational and psychological factors.

The need to involve the human factors and ergonomics specialist in the production phase of systems development seems clear. Helander and Domas[4] described a systems approach to analyze human-robot interaction for task allocation. They discussed the traditional people versus machine comparison lists but pointed out that once tasks have been allocated between humans and robots, the human tasks should be evaluated with respect to job satisfaction. Product design and workplace design influence job characteristics. Thus, products and workplaces must be designed so that job satisfaction does not suffer.

A complementary approach, which combines the best of people and machines, should achieve optimum results. It is difficult, however, to include people issues in early phases of the design process. One phase that is overlooked most often is the production phase. Perhaps this is a carry over from an Industrial

Revolution tradition of "managers manage, engineers engineer, and workers work."

Meister[5], an advocate for the role of human factors in systems development, asserted that the production phase is of little concern. He argued that human factors efforts are required in at least five of the six major phases of system development: planning, pre-design, detail design, test and evaluation, and operations, but not necessarily production. According to Meister, the questions to be asked during production are those of the industrial engineer and, therefore, do not concern us here.

The authors of this chapter disagree. The industrial engineer must consider human factors and ergonomics in the manufacturing design process. The following case study of an IBM manufacturing facility effectively illustrates the benefits of considering these factors in the design of manufacturing processes, even when the design criteria are relatively inflexible.

ANALYSIS AT MANUFACTURING PLANT

The manufacturing process described in this case study uses leading-edge technology in today's transitional period between products, which can be partially assembled automatically, and products of the future, which can be produced completely by machine. This process has been recently implemented and is presently undergoing final check-out (Figure 19-1).

All parts to be assembled are automatically delivered to either IBM 7547 robots or manual work stations. Figure 19-2 illustrates a robot assembling two parts, and Figure 19-3 shows an operator placing a product on a carrier. There are 18 production work stations with the robotic work stations making up 70% of the process. These work stations are interspersed with the remaining 30% of the process, which are manual work stations. These percentages can change with different products providing flexibility for future designs. Twenty additional manual work stations exist for audit and back-up, see Figure 19-1, and alternate facilities exist if the robotic work stations become incapacitated. After assembly and test, the product automatically leaves the process for shipment. In this assembly there is no manual materials handling such as carrying or pushing, although these activities still pervade a majority of presumably automated processes.

The current product design incorporates some design for automation (DFA) concepts, design modification that makes it easier to assemble the product by using automation. For example, only one type of screw fastener is used throughout assembly of the product. In other details, the design was essentially based on design efficiency and marketplace requirements.

Task allocations were based on what the robots could do within the constraints of the total system, with the remainder of the tasks given to people. People tasks were those that were either difficult to automate, like cable routing, or those that were too costly to automate and/or would disrupt

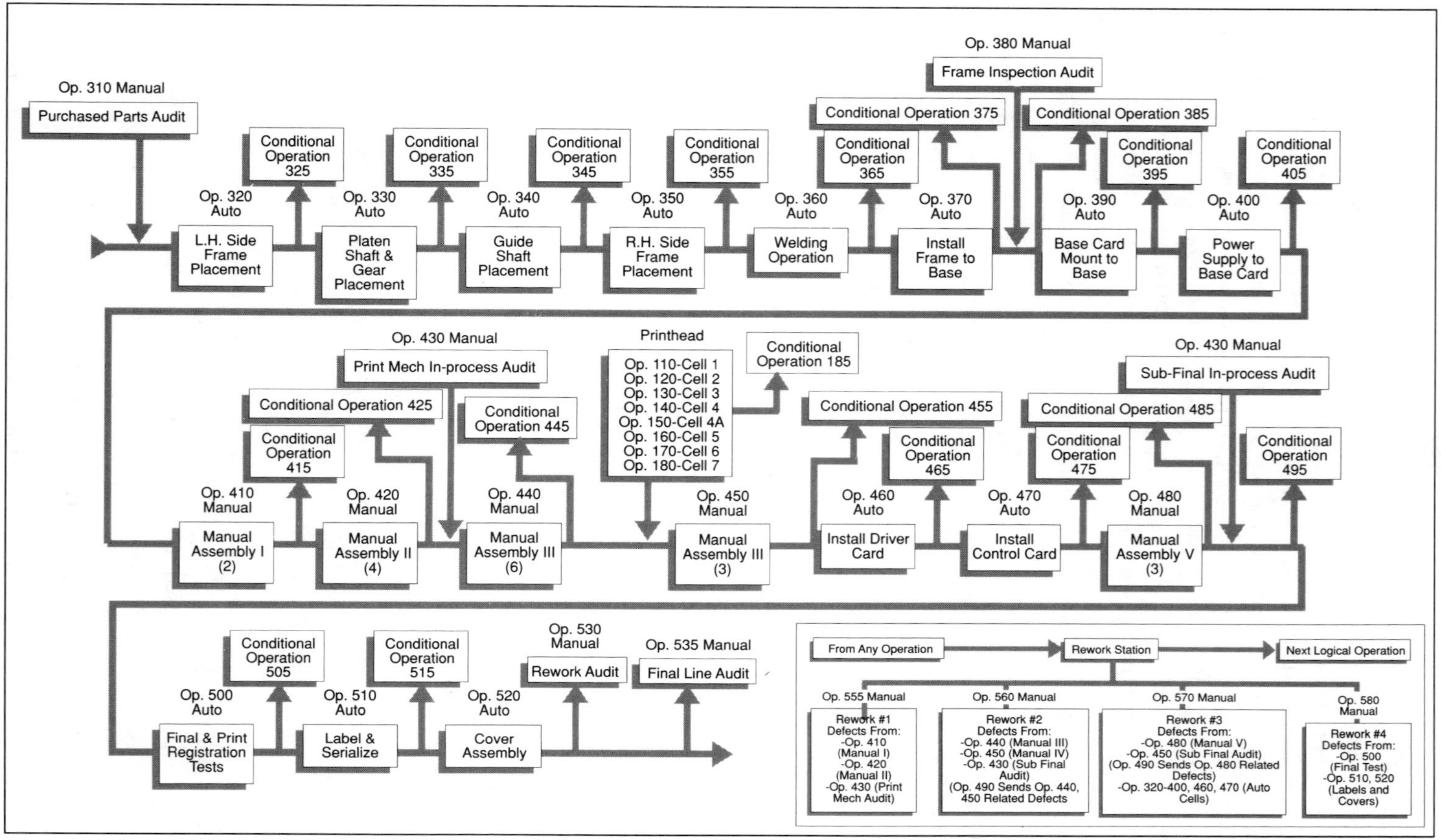

Figure 19-1. Illustration of 18 Production Work Stations Along This Assembly Line (Note: The manual work stations have been concentrated so that the operations can communicate. In addition, there are four manual work stations for rework.)

Figure 19-2. Use of a Robot for Automatic Assembly.

Figure 19-3. Use of a Roller Ball Transfer to Ease Manual Handling of Products.

scheduling. The manufacturing process, for the most part, had to be designed to fit the product. The challenge was to design a process compatible with people that produced a high-quality, high-volume product at low cost.

During our review we observed that the entire manufacturing team was aware of the importance of human factors/ergonomics. We attributed this to a combination of expressed IBM corporate interest and location emphasis fostered by ergonomics training programs during the last five years. For example, a manufacturing engineer responsible for equipment design has been incorporating human factors principles in design ever since receiving IBM human factors training in 1980. Without it, he stated, otherwise excellent designs may have provided only mediocre results.[6]

We have grouped the findings of our evaluation into three areas of consideration: task, environment, and tools. Each area discusses what was done to improve the interface with people and robots, the rationale, and the relevant research.

Task Considerations

The tasks in the process were simplified to accommodate the robots, and the resultant 30 second cycle time per operation was the basis for cost justification of the machines. For the human operator, however, such short-duration tasks might have led to boredom, fatigue, and injuries.[7] This potential problem was solved by designing into the process multiple manual work stations that provided the human operator with a minimum cycle time of two-and-one-half minutes. Some operations were increased to six minutes.

Table 19-1. Task Design Improvement Measures.

•	Increased assembly cycle times from 30 seconds to a minimum of 2.5 minutes
•	Recirculated conveyor system to reduce operator pacing
•	Developed flexible schedule for rest breaks
•	Initiated job rotation--two jobs per day
•	Changed physical proximity to enhance verbal communication
•	Established path for operators
•	Created buddy system for robotics safety

The difficulties associated with machine pacing of operators was humorously depicted by Charlie Chaplin as early as 1936 in the motion picture "Modern Times," and subsequently have been well-documented in research.[7] To avoid pacing of the operator at this assembly line, there is a recirculating conveyor system with work-in-process warehousing capability. The operator then can leave the workstation for a rest break without adversely affecting the system flow.

Job rotation is a technique that has been widely used to reduce boredom on the shop floor. In this facility, operators change work stations twice a day in reverse order to the direction of process flow. In this way they will not be checking their own work from the previous operation. Before a person can operate in a work station, the master system computer must verify that the operator is trained and has performed the task during the last 30 days. If this is true, the computer terminal in the work station accepts the operator and unlocks the station.

Perhaps the most important consideration in the people and machine interface is safety. This facility uses the "buddy system," which is analogous to the electrical safety rule of "never work alone." In other words, for all work done with robots there are always two individuals, one watching the person who is interacting with the robot.

Teamwork between all people involved—operators, engineers, and management—is essential to the performance of the project. Close physical proximity of the people involved aids in social interactions and clear communications on which teamwork can be built. With this in mind, the line managers' offices were relocated from a remote area to 20 feet from the manufacturing line. Support engineers' offices are located immediately across the hall.

Task performance data of individual operators is supplied by a computer to managers and operators for evaluation, recognition, and corrective action. Operator job categories have been defined, for example, assemblers, setup operators, quality and process auditors, and automated equipment operators. Task performance input and well-defined job categories aid career path discussions between the manager and operator, and help reinforce teamwork.

Environmental Considerations

The environment in which the activity takes place is as important as job content. At the IBM facility, special efforts are taken to provide the human operator with an efficient and comfortable workplace (Table 2). Work surface heights are electrically adjustable to fit both a tall and short operator. In addition, the operator can easily readjust the height throughout the course of the day. Vacuum suction cups on the work surface hold the product in different positions. Part bins are movable and can be grouped within easy reach of the operator. Roller ball transfers move the product to and from the work station. New chairs have been introduced that allow the operator to either sit or stand in the workplace with a simple adjustment.

Each operator's work station can be personalized by using their own tackboard. Lockable personal storage is also provided to promote a sense of security and to help personalize and organize the work station. These factors along with the plan to allow the use of music headsets by production operators, added to a positive work environment.

Table 19-2. Environmental Design Improvements.

• Work surface height electrically adjustable
• Vacuum suction cups to hold unit in different positions
• Roller ball transfer to aid manual handling
• Ergonomic chairs for sitting or standing work posture
• Tackboards for personalization of work station
• Lockable personal storage
• Use of low-noise equipment
• Enclosure of noise sources
• 1000 lux ambiant illumination
• Optional task illumination
• Adequate heat and ventilation for comfort

Many researchers have reported the adverse effects of noise on job satisfaction. With the high degree of mechanization in the plant's production area, the potential for a high noise level was great. Consequently, special attention was given to procure equipment that generated low noise levels or, where possible, to enclose the sound at the source. In one instance, noise was practically eliminated by changing frame serialization from stamping to laser etching.

Another important environmental factor is adequate lighting. For general assembly work at IBM, the ambient illuminance is usually about 1000 lux. Precision tasks need more illumination and coarse tasks may need less illumination. At this plant, additional task lights for human operator's are optional. Finally, the heating and ventilation requirements were established in accordance with IBM facilities practices to ensure a comfortable work environment.

Tool Considerations

The power tools used in this operation are lightweight and more delicate in appearance than the heavy-duty power tools typically used in the manufacturing environment. This increases efficiency and reduces the chance of repetitive motion trauma.[8] Some considerations of tool design to enhance ergonomics are listed in Table 19-3.

Table 19-3. Tool Design Considerations to Enhance Ergonomics.

• Light-weight power tools
• Reduction of number of controls on robot panel
• Schematic lines on panels to identify control sequences
• Use of color coding and color contrast
• Light fixtures inside control panels to simplify maintenance

The number of controls on the robot panel were reduced and schematic lines identifying control sequences were added to make operation easier to understand. The controls were grouped logically and conveniently laid out, with different groups having contrasting colors, which helps with legibility. These measures should reduce the cost of training, and the chance of human error.[9]

Maintenance is another area in which simple enhancements can provide dramatic improvements and reduce long-term costs. One common problem in maintenance activities is the lack of adequate lighting. Extension cord lights and flashlights are often used, which can result in inadequate illumination or safety hazards. To alleviate this problem, permanent lighting fixtures were installed inside equipment panels.

Evaluation of Work Stations

Since the installation of this manufacturing facility, injuries, strains, and complaints have decreased markedly. Operator performance at work stations has been virtually trouble free, and all of the manager's and operator's attention can be directed toward production rather than toward problems and injuries.

Operators appreciate the ergonomic design of the work station, the flexibility of the adjustable work surfaces, and the convenient positioning of parts. Managers like the flexibility of the work station. Finally, medical doctors at this plant indicate that the use of the ergonomic work station has eliminated most complaints of back and shoulder strain. This work station design has now become the plant standard and will be used for electronic card assembly in the future.

DISCUSSION

Safety and other performance outcomes should be the main criteria for systems evaluation. There is a tendency, however, to pay undue attention to irrelevant system behavior at the expense of performance. For example, the behavior of a traditionally trained army can be measured by assessing their discipline, but combat performance may suffer. Bailey[9] argued that one should focus on performance and the outcome, not behavior.

The challenge is to identify important performance criteria and how they interact with other (less crucial) aspects of human behavior. Of course some types of behavior are likely to remain important, although they are only indirectly associated with performance. For example, the design team that encourages creative human behavior may very well achieve performance excellence as a by-product. At the same time, job satisfaction is usually enhanced. A similar problem can be seen in architectural design that enhances aesthetics. Human design criteria for buildings, processes, and equipment often trades off in favor of architectural beauty. In the IBM manufacturing system, thought was given to appearance, but the primary design criteria were production efficiency, safety, and job satisfaction.

Process design on the work done by people and robots should ideally be evaluated when the product to be manufactured is designed. This is a very important stage, since the design virtually dictates the assembly process and the types of tasks available for individuals and automated processes. Procedures should be established that describe complementary activities for people and robots and should focus on achieving optimum production. Insufficient attention to human factors and ergonomics during the production design phase should be seen as counterproductive.

In the case of the IBM manufacturing facility, we concluded that without early involvement in the system design phase and an overall awareness of human factors and ergonomics many of the improvements would not have happened and the outcomes would have been sub-optimum. The early tests of this production facility disclose superior product quality and reliability. As manufacturing builds up to capacity, management is confident that expectations will be exceeded, not only for production but for people as well. Hopefully, the discussion and examples in this chapter will increase enthusiasm for the future benefit of both products and people by considering human factors and ergonomics early in the design phase.

References

1. Jordan, and Nehemiah. Allocation of functions between man and machines in automated systems. *Journal of Applied Psychology*. 1963:47(3):161-165.

2. Gerwin D, and Leung TK. The organizational impacts of flexible manufacturing systems: some initial findings. *Human Systems Management*. 1980;237-246.

3. Nof S, Knight, Jr. JL, and Salvendy G. The utilization of industrial and skills analysis approach. *AIIE Transactions*. 1980:12(3):216.225.

4. Helander M, and Domas K. Task allocation between humans and robots in manufacturing. *Material Flow*. 1985;3:175-105.

5. Meister D. The role of human factors in system development. *Applied Ergonomics*. 1982;13(2):119-124.

6. *IBM Ergonomics Handbook*. Armonk, NY: International Business Machines; 1981;53-59. Publication SV04-0224-01.

7. Salvendy G, and Smith MJ. *Machine Pacing and Occupational Stress*. London: Taylor and Francis Ltd.; 1981.

8. Hasselquist R. Increasing manufacturing productivity using human factors principles. *Proceedings of the 25th Annual Meeting of the Society*. Santa Monica, CA: The Human Factors Society; 1981.

9. Bailey R. *Human Performance Engineering. A Guide for Systems Designers*. Englewood Cliffs, NJ: Prentice-Hall, Inc.; 1982.

George J. Burri, Jr. is an Advisory Engineer at IBM Corporation Systems Integration Division, Federal Systems, Boulder, Colorado. He specializes in human factors engineering for aerospace projects and workplace ergonomics. He obtained a BS degree in Ergonomics & Business through the University Without Walls from Loretto Heights College, Denver, Colorado. He is a senior member of the Institute of Industrial Engineers, Society of Manufacturing Engineers, and a member of the Human Factors Society.

Martin Helander is an Associate Professor of Industrial Engineering at State University of New York at Buffalo. He specializes in Human Factors Engineering and his research interests include human-computer interaction, industrial ergonomics, and driver performance. Dr. Helander obtained a PhD in Engineering at Chalmers University of Technology in Goteborg, Sweden, and is a Docent of Engineering Psychology at Lulea University. He is a fellow of the Human Factors Society and The Ergonomics Society.

Section V

Product Design

20

Human Factors in Product Design

Jerry R. Duncan

BASIC PRINCIPLES

The concept of the human-machine system is the basis for human factors applications in the product or system design process. This concept or model states that there is a closed-loop relationship between workers and their equipment (Figure 20-1). The operator of a corn harvesting combine, for example, receives information from the equipment through displays in the cab or from the external environment. This information may include the position of the picking units with respect to the corn rows or the amount of grain on the ground behind the combine. The operator processes the information and makes certain decisions, such as turning the steering wheel or reducing the combine's ground speed to achieve some desired goal. The equipment consequently provides new information (feedback) about its changed status to the operator, and the cycle continues. This interaction between the operator and the machine creates the human-machine system relationship.

Human factors design philosophy contains four basic principles,[1] which are key to successfully incorporating human factors considerations in product or system design. The first principle states that the functional effectiveness of equipment is related to the efficiency with which people can operate and maintain the equipment. Thus, a decline in the operator's performance may cause the equipment to perform less effectively or fail to perform its intended function. Although a self-propelled forage harvester may have the capability of traveling at 15 kph through a field, for instance, it will fail to perform at that level of productivity if the operator feels uncomfortable traveling at that speed and selects a slower, more comfortable ride.

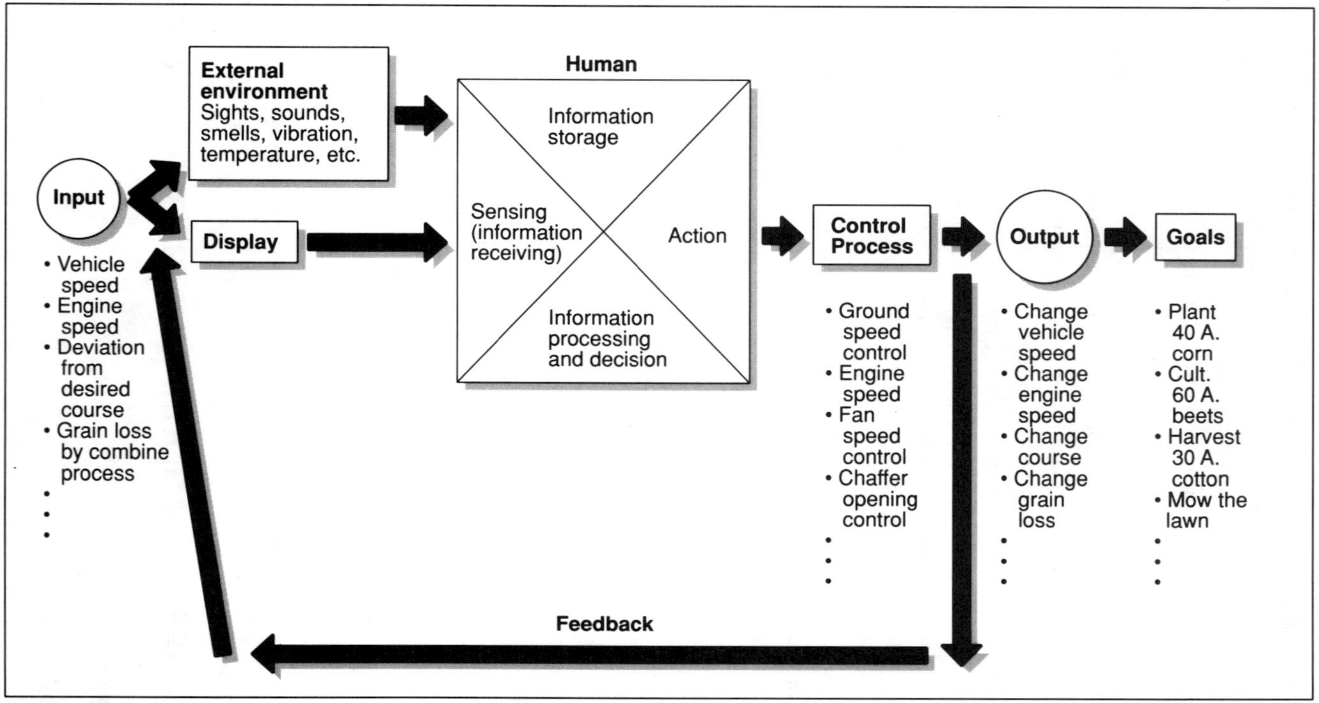

Figure 20-1. A Model of a Human-machine System.

The second principle claims that equipment design influences how people operate and maintain equipment. Characteristics of the equipment design, such as the responsiveness of the steering system or brakes, the visibility of information displays in sunlight, or the force required to loosen a drain plug, act as stimuli to which the operators must respond. If the equipment characteristics (stimuli) require more from the operators than they are capable of providing, they may respond incorrectly, out of sequence, or not at all.

These equipment stimuli also demand selective attention of the operators. Operators cannot respond to all of them at the same time. At any given time, the operators must pay more attention to some features of the equipment than to others. These stimuli can also take the form of procedures for operating and maintaining the equipment, which can often impose significant loads on the operator's memory.

The third principle is derived from the second. Since equipment design characteristics act as stimuli to the operators, then it follows that operators will respond more efficiently to certain arrangements of these characteristics over other arrangements. If equipment characteristics are designed to match the capabilities and limitations of the operators, the operator's performance should be more efficient. For example, whole body resonances that occur in the frequency ranges between 3 Hz and 6 Hz and between 10 Hz and 14 Hz are uncomfortable to the operator. So, a seat or cab suspension design that greatly limits a vehicle operator's acceleration in these frequency ranges will be better than one that does not.

The fourth principle states it is easier to modify equipment characteristics to match human capabilities than it is to modify (or select) human capabilities to accommodate the equipment characteristics. It is easier, for instance, to select displays with larger alphanumeric characters or to arrange displays so that they are closer to the operators than it is to add more sensitive visual acuity to the operators. It is also easier to design controls to be actuated within human force capabilities and to locate them within human limits of reach than it is to endow the operators with greater strength or to change their physical dimensions to a more desirable geometry.

A PRODUCT DESIGN PROCESS

In highly competitive industries, meeting scheduled introduction target dates is crucial. Consequently, effective product development requires that a well-planned design sequence be followed (Table 20-1). Product engineers are responsible for organizing the development of an idea into a profitable product, which satisfies a specific human need.

To be successful, the product must have a superior appearance, meet performance standards, be highly reliable, be priced to sell, and be profitable. Economic studies, such as market surveys, estimated engineering and development costs, capital expenditure costs, and unit cost are essential to deciding whether or not to accept a product design. This information also helps determine the anticipated selling price and potential profits from the product.

Table 20-1. A Product Development Sequence.

1.	Identify a human need.
2.	Create product ideas to satisfy human need.
3.	Evaluate alternative design ideas.
4.	Design and build prototypes.
5.	Test and evaluate prototypes.
6.	Perform "pilot" run of manufacturing process.
7.	Implement full-scale manufacturing.
8.	Conduct product "follow-up."

Another important step in the product design process is to develop and evaluate alternative designs prior to the product's introduction. The evaluation of design ideas may include a functional requirements review, during which the soundness of the design is scrutinized, impractical ideas are eliminated, and a check is made to ensure that original function requirements are satisfied.

The evaluation may also include design group reviews and analytical methods. Design group reviews often find potential deficiencies in the design or considerations that have been overlooked. The analytical methods, including mathematical models and computer simulations, provide information about stress within the structure for a given load condition, fatigue life of components, and how parts fit together within the design. These methods are valuable, because they permit product engineers to consider design changes and to evaluate the effects of those changes, with considerably less time and expense compared with methods requiring physical testing.

Once the design is satisfactorily reviewed, a general product specification is developed. Often, this is a joint effort between the marketing and product engineering departments. Product specifications help ensure that the product has well-defined specifications and is acceptable to all appropriate departments of the manufacturing firm.

Many design selection criteria operate simultaneously during the design process, each of which is important. Many of these criteria can conflict, making it impossible to maximize all of them. The design selection criteria listed in Table 20-2 are representative of those that are generally balanced against each other.

Once a product design has been selected and prototype units built, the next step in the process is testing and evaluation. A thorough test and evaluation program requires a major commitment in manpower, facilities, and time. It generally includes activities of component testing, field stress testing, durability testing, and a complete system evaluation. During this period, further safety reviews and human factors evaluations are conducted.

Table 20-2. Design Selection Criteria.

•Functional Performance - Speed - Accuracy - Maintainability - Serviceability - Operability - Reliability - Safety • Compliance to Standards and Regulations - ANSI - ASAE - OSHA - ISO - SAE - Corporate standards • Styling • Cost - Cost to design - Cost to manufacture - Cost to service • Others

HUMAN FACTORS CONSIDERATIONS

Human factors variables are considered throughout the product design process. At the beginning, requirements for the operator are usually specified. These requirements include space, environmental conditions, acceptable control forces, and others. The system's functions are described, and choices are made as to the allocation of functions to either human or machine control.

The functions to be performed by the operator may then be subjected to a detailed task and operations analysis. These analyses identify and resolve potential conflicts with the specified requirements. The operator's work space, including the seating, controls, displays, and climate, is then designed to best accommodate the specified requirements. Training materials (operator's manuals) must also be designed to be compatible with the machine design and the operator's needs. Testing and evaluating the operator's workstation may be conducted in laboratories, during which performance indicators, for example, speed, accuracy, preferences, and errors, can be measured in controlled, repeatable conditions or in field test sites when realistic environmental conditions are desired.

Human factors specialists are generally part of a team that supports product engineers. Because the product engineers put great value on the timeliness of the response, the available amount of time determines how extensive an effort can be made for gathering information, analyzing data, and reporting recommendations. Sometimes the specialists are asked for some needed information just a few days before a final design decision will be made. When this occurs the adequacy of the response will largely depend on the tools available at the time. These tools can include the telephone to call an expert colleague; a library of technical literature; computer databases, either privately developed to meet unique needs or commercially available general ones; laboratory facilities for conducting evaluations of product concepts by physical simulation; observing and recording the behavior of the prototype's users; or final products at selected field test sites.[2, 3]

Human factors specialists are catalysts for change; they identify improvements to existing products or proposed designs. The proposed changes generally address improving some measure of human performance or comfort. As such, their goals may often conflict with those not wanting to change an existing product design or a manufacturing process. An inertia often protects an incumbent product design from challengers. Proposed design changes or new constraints created by other disciplines also compete with the goals of human factors specialists.

An executive commanding the production engineering organization at a Detroit-based manufacturer, for example, wanted only engineers who were willing to design things that the current production machinery could make. The objective was to reduce the cost of manufacturing a product. A possible consequence of this objective, and one which a human factors specialists and perhaps others would challenge, is that a hole for locating a control lever would be punched where a production machine can do it at lowest cost rather than where the product user needed the lever.

CASE STUDIES

The following are some examples of how the human factors engineering discipline and principles have been applied to the design of off-road vehicles and outdoor power equipment.

Case Study One
Transmission Shift Quality

As noted,the human factors specialists' contribution to a product design is greatly influenced by the time and tools available to answer a given design question. When asked to provide information about how tractor operators perceive the acceleration experienced during a transmission shift from one gear to another, the needed research tools and lead-time were available.

The specific design question was: How do the parameters of peak acceleration, peak jerk, direction of initial acceleration, and background vehicle vibration affect "acceptability" of tractor transmission shifts? To answer this question, the specialists needed a unique research tool, and a vehicle simulator (Figure 20-2), which Duncan and Wegscheid[4] described in detail. This study also required the use of a rating scale, a more common human facotrs research tool.

Volunteer operators steered the tractor simulator continuously along a computer-generated roadway projected onto a screen in front of the cab. At selected intervals the operators were given a visual cue to move the transmission shift lever. Movement of the shift lever caused a transient motion in the fore-aft direction superimposed on a background vibration for the tractor simulator. Movement of the steering wheel caused the computer-generated visual scene to appear as though the operators were driving a tractor along a winding road. Following each shift the operators reviewed a rating scale placed in front of them, judged the acceptability of the shift, and spoke into a microphone a number to characterize the acceptability of the shift.

On the rating scale, a “1” indicated a “totally unacceptable” or bad shift, and a “9” indicated a “totally acceptable” or good shift. The two levels of background vehicle vibration investigated in this study were composed of two simultaneous motions: a vertical motion and a roll motion. The vertical motion was based on an International Standards Organization document that describes the acceleration power spectral density to be used for Class II tractor. Maximum roll was +- 2°. The lower level of background vibration was set at approximately 28% of the high background level for both vertical and roll motions.

Figure 20-3 shows the relationship between peak acceleration and vehicle background vibration on the operators' perception of “bad” shifts. The specialists defined a bad shift as one receiving a rating of 1, 2, or 3. Figure 20-3 also shows the percent of the total observed bad shifts when the peak jerk was 7 g/s at the low and high levels of background vehicle vibration. The percentage of bad ratings increased rapidly as peak acceleration increased between 0.4g and 0.7g. Nearly all operators agreed that above approximately 0.7g any shift was bad. Figure 20-3 also shows that increasing the vehicle background vibration during a shift decreased the percent of bad shift responses at a given peak acceleration value. Information gained from this research allowed the transmission engineers to make design judgements that accommodated the tractor operators' expectations of shift quality.

Figure 20-2. Vehicle Operations Simulator at the Deere & Company Technical Center's Human Factors Laboratory.

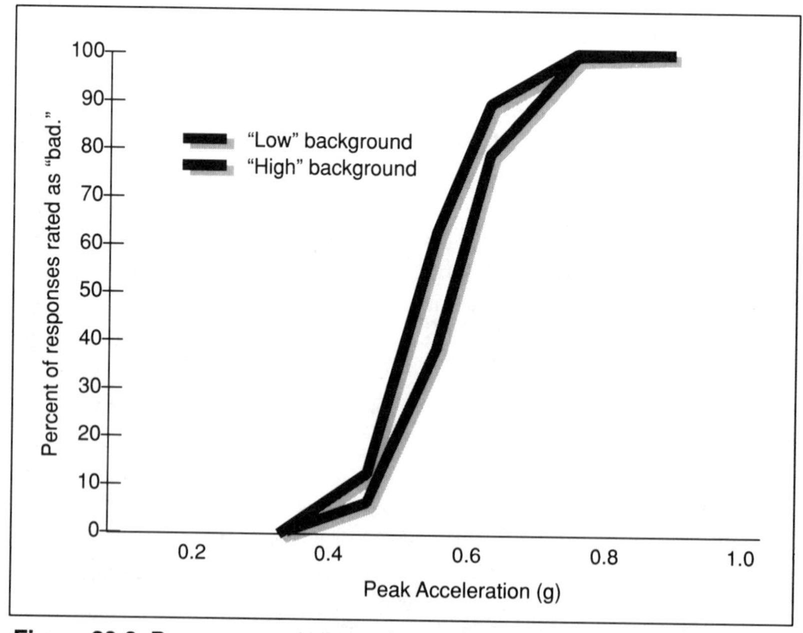

Figure 20-3. Percentage of "Bad" Shifts at 7 g/s Peak Jerk for Low and High Levels of Background Vibration for Tractor.

Case Study Two
Corn Harvesting Combine Automation

Product engineers wanted to know how operators of corn harvesting combines would react to combine automation. The vehicle simulator was used in this study to simulate the combine, and experienced combine operators were used as subjects. After the simulator test sessions, operators were interviewed and asked to complete questionnaires about the session. As a result of this study, the engineers learned how automating combine functions affected system performance and how operators adapted to the changes. Turner et al[5] describe the development of the combine simulator.

Case Study Three
Off-Road Vehicle Speed and Direction Controls

Following an evaluation of commercial lawn mowing equipment in a field environment, a product engineer asked for a review of various configurations of foot and hand controls used for controlling the mower's ground speed and direction. The vehicle simulation facility was used in order to control the necessary environmental and task variables. The objective of the study was to determine how the use of hand levers and foot pedals affected productivity and operator errors.

Using a simulated vehicle operator's workstation, 20 operators performed a material handling task, simulating the repetitive fore-and-aft travel of a vehicle with a front-end loader. The simulator allowed easy alterations of the hand and foot control configurations. Productivity, as determined by the time to move material from one location to another, and various operator errors, such as activating the wrong direction control, were recorded. After analyzing the results of the initial phase of this research, in which five different control configurations were evaluated, a sixth arrangement was devised, which was expected to out-perform the others.

One of the control configurations tested was similar to the conventional speed and direction controls on an industrial loader or forklift truck. That is, direction (forward and reverse) was selected with a left-hand lever at the steering wheel rim, and speed was controlled with an accelerator pedal operated by the right-foot.

Figure 20-4 shows the performance of 20 operators using the conventional controls compared with the performance of 15 of the same operators using the new two-pedal configuration. With the new configuration, the operator controlled forward speed with the accelerator pedal and reverse speed with a short pedal located to the right and slightly to the rear of the accelerator. Shuttling between forward and reverse was performed by pivoting the right foot around a heel-point and pressing forward to go forward and pressing downward and rearward for reverse.

The graph shows that of the 267 cycles (loading, dumping, and loading again) included in the analysis for the conventional configuration, 50% of the cycles were completed in less than 57 seconds. Of the 195 cycles included in the analysis for the new configuration, 50% were completed in less than 45 seconds. The new configuration reduced the cycle time by 21%.

When compared with the previous best configuration, a dual-pedal configuration in which the operator pushed forward on a pedal to go forward and pushed forward on another pedal to go in reverse, the new configuration was 8% better in terms of productivity (shorter cycle time), and contributed to a significant reduction in the number of operator control errors.[6]

In an attempt to persuade the product engineers to strongly consider this new control concept, a small tractor was modified to demonstrate the new control configuration. As a result of providing the product engineers with a working prototype, the control concept derived from the human factors research led to the rapid incorporation of the new control into a commercial product. A detailed description of the development of the John Deere residential front-mower, which uses the new control, is given by Flenniken et al.[7]

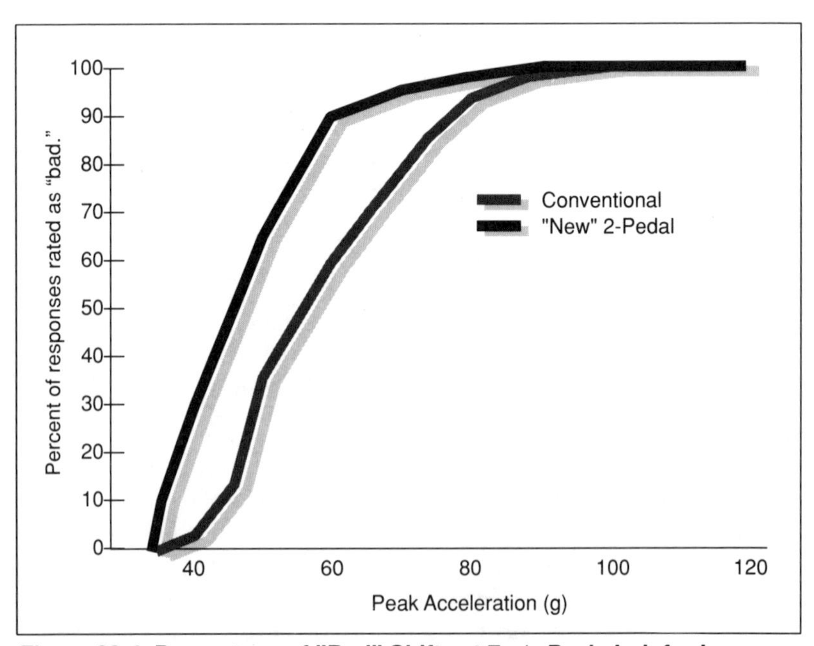

Figure 20-4. Percentage of "Bad" Shifts at 7 g/s Peak Jerk for Low and High Levels of Background Vibration for Off-road Vehicle.

CONCLUSION

The following are some of the basic facts or givens regarding human performance that can be used to guide product design, or at least identify design questions that deserve attention:

- People are different in their size and shape (anthropometric dimensions). Because of this fact, the human factors design philosophy leads the design of a vehicle seat, for example, to permit the operator to independently adjust the vertical and fore-aft position. This allows the operators to determine their viewing position for reading displays and observing the tasks, and choosing how they want to position themselves relative to the machine's controls.
- People are different in their perception of comfort. This fact motivates designers to provide adjustability in the operator's work environment (seating, foot/leg room, and air conditioning) to accommodate these differences.
- People are different in their motor capabilities such as reaction time and strength.
- People are different in their sensory capabilities such as visual acuity, hearing ability, and smell sensitivity.
- People are different in their mental or cognitive capabilities for storing and processing information and for making decisions. The rate at which they learn, the ability to retain facts in memory, and the ability to retain skills are different.
- People are different in their experiences, native-training, motivation, cultural background, perception of risks, and many other characteristics. Even the same person may be different at different times.
- People can be observed by the way in which they compensate for unwanted system characteristics and behavior. Compensating behavior, that is, human actions in response to interferences to achieving a goal can be observed in many settings. A common example one is the use of personal computers. In the situation where a computer user is trying to create a simple XY graph, using a widely-used spreadsheet software product, the user often "gives in" to the product and accepts the graph in the form that the product constrained it. This compensating behavior generally stems from the user having expectations of the product that will not be fulfilled, such as, believing that the computer graphing program is as accommodating as pencil and paper. Although people are adaptable, their adaptability can compensate only to a limited degree before system performance is degraded or before the user becomes dissatisfied and either tolerates the product as it is or searches for another alternative.
- People are goal oriented when operating machines or when performing as system elements. They make mistakes, and they learn from their experiences.

- Somewhere in the general sequence of performing a task an operator may be faced with choosing among alternative goals or perhaps coordinating conflicting goals. For example, while using a product the goal of completing a task may conflict with the goal of avoiding compromises in safety.

  A case where human behavior, directed by a goal, has affected the performance of a safety device concerns the removal of tractor *pto-shields*. A *pto* is a rapidly rotating drive shaft called a *power take-off* that is used to mechanically transfer power from a tractor to an implement attached to the tractor. An article in *Successful Farming*[8] summarized a survey conducted by Purdue University in four midwestern states, which found that the pto-shield is more likely to be left in place if it is of the flip-up type or hinged at the back so it lifts up out of the way and then flips back down. Flip-up shields were found in place 89% of the time. Bolted-on, fixed shields were found in place 58% of the time, and quick-attach shields (easily removed) were found in place only 31% of the time. They found tractor operators removing the shields because of interference with the use of centrifugal spray pumps, interference with fully mounted equipment such as posthole diggers, and interference with tasks of connecting and disconnecting drivelines. These data demonstrate that the equipment design does influence the operator's behavior.
- People are agents of errors. Although they may seem to be a prominent and persistent source of error, they generally propagate more than cause error-likely conditions. An important element in the system design process is to analyze the tasks performed by the equipment operator and to classify potential human error associated with those tasks. This analysis will likely suggest possible equipment design or operator training remedies to either reduce the frequency of errors or to make the system more error tolerant.
- People will accept or take risks to achieve goals. Their experience affects their current decisions. The operator's application of knowledge from one system to another has led system designers to utilize population stereotypes to minimize operator training time and to reduce the likelihood of negative transfer of training.

References

1. Meister D. *Human Factors: Theory and Practice.* New York: John Wiley & Sons; 1971.

2. Meister D. *Behavioral Analysis and Measurement Methods.* New York: John Wiley & Sons; 1985.

3. Meister D. *Human Factors Testing and Evaluation.* Amsterdam: Elsevier; 1986.

4. Duncan JR and Wegscheid EL. Off-road vehicle simulation for human factors research. *ASAE Paper*; 1982. 82-1610.

5. Turner RJ, Duncan JR, and Wegscheid EL. The development of a corn harvesting combine simulator. *ASAE Paper*; 1985. 85-1578.

6. Duncan JR. Design of off-road vehicle speed and direction controls. *ASAE Paper*; 1989. 89-1112.

7. Flenniken JM, Swanson LD, Hardesty LR, and Hardzinski JE. Development

program for the F510/525 residential front mowers. *ASAE Paper*, 1989. 89-1583.
8. Finck C. The great pto-shield rip-off. *Successful Farming*, March 1986.

Jerry R. Duncan is a human factors specialist at Deere & Company Technical Center, Moline, Illinois.

21

Human Factors and the Design of A Data Acquisitions Van

Steven M. Casey, PhD

Hydraulic fracturing is an advanced technique for recovering oil and gas from deep underground wells. The process involves pumping a dense mixture of water, sand, and chemicals down a well at a very high pressure. The fracturing treatment requires as many as 40 vehicles on-site, including diesel or jet turbine pumping trucks; chemical tanks; water tanks; sand trucks; sand-handling equipment; a vehicle to blend the sand, water, and chemicals; and a large bus from which operators control the process, called a *frac bus*.

Sand, water, and chemicals are blended throughout the treatment in prescribed proportions and pumped at pressures between 5,000 psi and 20,000 psi (Figure 21-1). The rate at which the geologic formation accepts the material and the pressures monitored at the surface indicate the width, height, and length of the fractures in the zone containing the oil or gas. By the end of a day-long treatment, 1 million to 2 million pounds of material may be pumped into the well, creating underground fractures and providing new channels of permeability to the well bore. Throughout the day, the supervisor, equipment operators, and customers require critical information to monitor the treatment and, if necessary, to modify the pumping parameters, pressure, rate, and slurry density.

The Treatment Monitor Vehicle (TMV) is a totally new concept in treatment data acquisition. Although data acquisition systems for well treatments have been in use for many years, a large digital system for acquiring, processing, and displaying treatment data was not in widespread use by any service company at the time the TMV was conceived. The goals of the system are to analyze and record data and present summary information to operators and customers during the treatment.

Figure 21-1. A Crew Prepares a High-pressure Manifold and Pumping Trucks for Hydraulic Fracturing Treatment on an Oil Well.

Such a significant departure from traditional systems required the design team to conduct research to ensure that the system can acquire the data needed by customers, present these data in comprehensible formats, and provide an appropriate physical setting from which the treatment could be monitored. The system also had to be relatively simple to assemble on site, easy to operate, and place as few new demands as possible on company training resources. Human factors research and recommendations were, therefore, needed to ensure that the system served its required function and was accepted by different types of users.

This chapter provides an overview of the human factors issues presented at the outset of the project, describes the human factors activities and subsequent recommendations made throughout the project's course, and summarizes an assessment of the final product.

OVERVIEW OF FUNCTIONAL REQUIREMENTS AND HUMAN FACTORS ISSUES

Functional Requirements of the TMV

This project began with a blank slate. Successful completion required that the functions of the proposed TMV system be defined in advance and that the information requirements of each of the system users be well understood.

Quality assurance. The TMV's primary function is to provide quality assurance data, with the basic goals of increasing the volume and quality of data available during and after a treatment and documenting the overall quality of the job. Based on the premise that the system is to provide data for monitoring the quality of service, three human factors questions were identified as being of special importance:

- Who are the individuals (customers) who would use the system, and what would be their level of technical sophistication regarding well treatments?
- What information is indicative of treatment quality, and what are the additional information requirements of the customers?
- Is there specific information that should not be presented in real-time in order to avoid the overall misinterpretation of job data and conclusions regarding treatment quality?

Process control. Fracturing treatments have traditionally been monitored and controlled through the Automatic Treatment Recorder (ATR). This analog device is usually located in a frac bus, and presents data on pumping pressure, rate, and slurry density. It has been the primary source of information for the service supervisor who orchestrates the actions of many individual equipment operators located in the frac bus. The chart record produced by the ATR has also served as the official record of the work performed during fracturing treatments.

The TMV was viewed, in part, as a vehicle for improving process control through the provision of more accurate treatment information. During fracturing treatments, for example, far more than the three basic variables of pressure, rate, and density could be monitored. More importantly, acquired data could be processed and summarized by computer systems. This tremendous improvement in presenting information could be used to increase overall control of the process.

This requirement involved resolving three questions, each of which has important human engineering implications:

- To what degree should the service supervisor rely on TMV data for control over the process?

- What TMV data will be required by the service supervisor to improve control over the treatment?
- How should the service supervisor acquire the data from the new data acquisition system? How could this information be made available without requiring the service supervisor to leave the frac bus and the ATR, the location from which the treatment is normally orchestrated?

Increased customer satisfaction. The design team recognized early in the project that introducing a sophisticated data processing unit to the field could serve to increase overall customer satisfaction by providing meaningful data to customers, and increase business by providing capabilities unique within the well service industry. Improved customer satisfaction, could result from presenting data of special interest and value, and from providing a more agreeable and functional environment from which to monitor a treatment and to evaluate the acquired data. This requirement would be attractive to customers contemplating large treatments, particularly where investments are high and the potential gains brought about by valuable treatment data would be substantial.

Many issues had to be addressed to ensure that the proposed system met this requirement, including:

- How many customers may be present at a treatment site, and how many of them should the TMV accommodate?
- What information would customers like to have access to during the course of a treatment, and what information should actually be presented to the customer?
- Besides the real-time treatment data, what other information requirements would the customer have during the course of a treatment?

Treatment data base. The proposed system could serve an additional function by acquiring, processing, and storing data for future reference by engineers and analysts. Such information might be used to study and improve standard operating procedures by providing an accurate record of a treatment, or improve subsequent treatments by examining the treatment record and the response of the well. This functional requirement involved the resolution of two issues:

- What well and treatment information should the system require for the purpose of supplemental data analysis by the company?
- Who is going to acquire and/or manage this information during the treatment, and how would these operations be accomplished?

Process control platform. Initially, the TMV was planned to serve as a data acquisition and processing center. In the future, the platform would probably be used for actual automated control of the process. This issue required defining the principal characteristics of a future, automated command and control center for the TMV?

Resolution of Functional Requirements

In order to successfully integrate the human factors effort into the development of the system, it was critical that these potential functional requirements be distilled into a reasonable set of requirements. Decisions were made regarding the role of the TMV and its potential users, and each requirement was weighted to reflected its relative importance. The following are the resulting TMV requirements:

- Serve as a system for acquiring and processing treatment data and presenting meaningful output to service customers. Customers might use this information to monitor the quality of the service, or they could use it to modify a treatment in progress. In any event, the TMV is a highly visible product, which would attract customers based on its unique capabilities and design.
- Provide information for quality assurance and process control. These features would not become required elements of the treatment service, however. Most treatments would continue to be run from the frac bus and the ATR log would continue to serve as the official record of the work performed. Service supervisors may occasionally monitor the job from the TMV, but the common control procedures would remain unchanged. The service supervisor viewed the TMV as a system provided primarily for the benefit of the customer. The information provided by the TMV would be available to the service supervisor if desired.
- Meet the immediate requirements and possibly serve as a platform for an automated control center at some future time. These additional requirements will not necessitate significant changes in layout or hardware. They will, however, require major changes in some equipment components and feedback loops.

Human Engineering Issues

Many human factors issues throughout the course of the project were addressed. These concerns have been organized into seven classifications:

- System users
- Information presentation
- Operating system design
- Interior design and layout
- Exterior work space design
- Operator selection
- Operator training

System users. Based on the functional requirements, an operator had to be present to configure and operate the system. One or more customers are likely to be present at all times, and one or more service supervisors may also be in the vehicle during a treatment. Other users may include district engineers and additional representatives of the customer such as geologists and petroleum engineers. The following questions summarize the basic issues regarding potential users of the system:

- How many users of each class (customers, service supervisors, district engineers) would likely need TMV data during a treatment? How many people should the TMV seat during normal operation?
- Should these individuals have direct control over the information presented on the display devices?
- What information should be provided to each of these groups of users?
- What will each of these individuals do with the information provided by the system?

Information presentation. Once the system users were identified, what information to display and how to present it could be determined. The human factors concerns in this area ranged from the actual content of the information provided to the characteristics of the systems presenting the information, for example:

- What classes of information, for example, graphics, tables, or digital readouts, will be displayed, and which display devices would meet these requirements?
- How many of these display devices should be provided, and how should the information be allocated to each device?
- How should information be formatted on each of the displays?

System operation. The TMV is operating system design is critical to the success of the vehicle in the field. The system could display a lot of valuable data to customers and service supervisors, but if the data were difficult to manipulate and manage, all individuals involved, including operators and primary users, would not welcome the introduction of the vehicle. The issues regarding the operating system generally pertained to the software with which the system was to operate. The primary concerns were:

- What characteristics of the operating software could ensure that an operator with limited experience with computers could allocate information to displays?
- How could the operating system/software be configured to minimize operator errors during input?

- How could the operating system/software be configured to minimize the time required to input data?
- How could the operator's task be simplified to minimize training requirements?
- What work aids or reference manuals would be necessary for configuring and operating the system?

Interior design and layout. The interior design and layout of the TMV is also very important to the ultimate success in the field. The vehicle had to be large enough to comfortably seat an operator and possibly four or more observers. The operator had to have ready access to all controls and displays of importance, yet could not be totally removed from the observers. The operator would also require a complete view of the treatment outside the TMV, and would need storage space for all materials that might be required. The observers would require a complete view of the treatment as well, and they would need to see the appropriate display, printer, and plotter. Seating had to be comfortable for all individuals, and floor space had to be adequate for movement in and out of the vehicle. These issues are summarized with the following questions:

- What tasks will the operator perform during each phase of TMV operation?
- Which displays and controls will the operator access during each phase of TMV operation?
- What is the best arrangement of displays and controls for operation?
- What are appropriate interior dimensions, including ceiling height; isle width, counter height and width; and seat height, depth, width, and back angle?
- What are the requirements for windows, and what are the appropriate dimensions of these windows?

Exterior work space design. The exterior design and layout of the TMV is as important to rapid and successful operation as the interior design and layout. Adequate storage for 30 to 40 cables had to be provided, and a system had to be developed for quick removal and storage of cables. Previous experience with research versions of a data acquisition vehicle had shown that a cable system was definitely required. Storage also had to be provided for each transducer carried by the TMV, including densitometers, flow meters, and pressure transducers. The densitometers are particularly important due to their weight and potential radioactive hazard. These and other issues are summarized as follows:

- How might cabling systems be designed to provide for quick access to cables, ease in cable placement, and rapid cable retrieval?
- How might transducer storage compartments be designed to provide for

easy and safe access to flow meters, pressure transducers, and radioactive densitometers?
- What systems might be developed, including computer aiding, to assist the operator when cables were to be strung throughout the treatment site?

Operator selection. Two general classes of employees may serve as operators of the TMV: company employees who are very knowledgeable in fracturing and possibly cementing services, and people acquainted with computer programming and data processing. Specifically, the necessary areas of expertise, skills, and knowledge, for example, computer processing or treatment procedures, of candidate TMV operators needed to be determined.

Operator training. Once operators were selected, each would be required to undergo training prior to operating the equipment. This training course had to reflect the basic abilities of the potential TMV operators and to instill the basic skills required to operate the TMV. The primary issues related to TMV operator training included:

- What information and skills must TMV operators possess after training to successfully configure, run, and interpret a TMV-monitored treatment?
- What is the appropriate design of a training course for TMV operators, and what exercises should be provided for students?
- What materials will be necessary for training a group of students on basic TMV operations?

CHRONOLOGY OF HUMAN FACTORS DESIGN CONTRIBUTIONS

Specified Functional Requirements

The basic functional requirements were addressed, by conducting structured interviews with future system users and by observing many fracturing treatments in oil-fields in Oklahoma, Texas, and California. During this early phase of the project, the objective was to narrow the lengthy list of design requirements into a list that was appropriate both functionally and economically. The following conclusions resulted from these initial analyses:

- The system would be located in a separate van or truck, and would not be incorporated into the frac bus, due to space requirements and the need for a relatively clean environment.
- The system would be provided primarily for the benefit of customers, although a company operator would have complete control over the data presented in real time.
- Data from the system would be available to service supervisors if required,

but the system would not officially replace the ATR as the primary record of the job.
- In addition to the operator, four or five customers or company employees should be accommodated by the system.
- The entire system must be far less cumbersome to configure and operate than data acquisition systems previously developed for research purposes.

Another major element of the functional requirements analysis was determining what was to be displayed and how it was to be displayed. An extensive list of information requirements was developed based on the structured interviews. The list addressed the customers, display (averages and trends), the operator and service supervisor's monitor (detailed process data), the x-y plotter (measured parameters versus time), the printer, and additional remote displays. Over 50 different display requirements and display elements were identified.

Developed Interior and Exterior Layout

The design team conducted basic design reviews relatively early in order to meet project deadlines and production schedules. Figure 21-2 illustrates the concept on which the design team agreed. Cables for all transducers would be located on cable reels located in compartments on the passenger side of the vehicle (Figure 21-3). Each cable would be clearly labeled, and could be deployed and stowed simply by winding the individual cable reel. The diesel generator would be located forward of the main compartment and behind the cab. This would reduce the noise level within the main compartment. All transducers would be stored in compartments on the driver's side of the vehicle, underneath the customer's couch. Heavy densitometers (168 lb each) would be stored on racks, which would roll out of the compartment for easy access (Figures 21-4 and 21-5). All other transducers would have dedicated storage compartments and racks. Some of the recommendations for specific features of the vehicle are outlined below.

Customer seating. The design team chose and designed bench-type seating to accommodate four people. Recommendations for seating dimensions were based primarily on MIL STD 1472. Retractable arm rests gave a greater long-term seating comfort and the capability for accommodating one reclining person.

Operator's work station. The operator sat on a swivel chair on wheels to enable movement around the work station while seated. The chair remained secured to the wall between the main equipment rack and customer seating area when the vehicle was in motion. It was not to be used during vehicle transport.

The operator's work space was designed to ensure easy access to all of the important components during the course of a treatment (Figure 21-6). The operator's display, on which all treatment- and operator-oriented information appeared, is an integral part of the computer. The densitometer panels (dis-

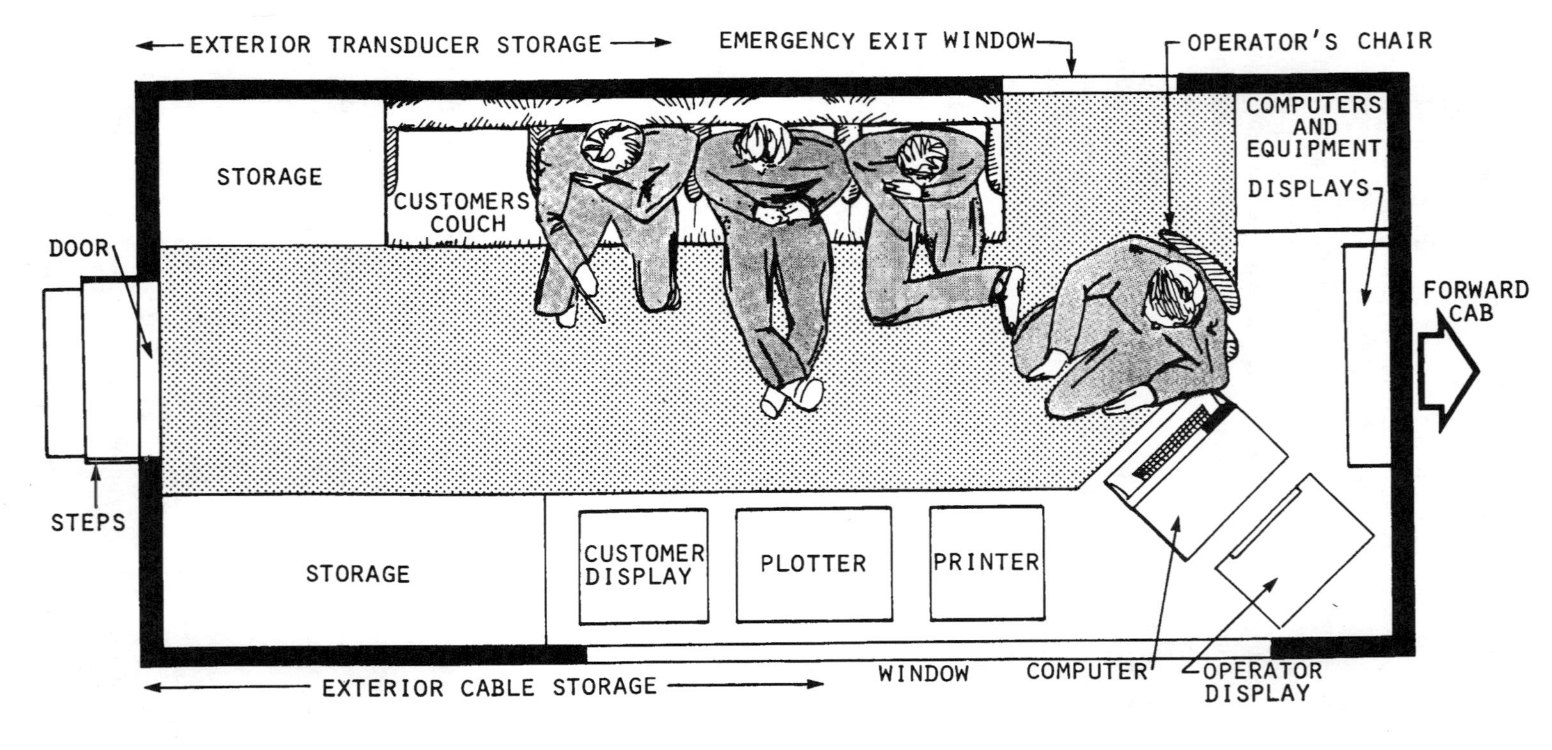

Figure 21-2. Plan View of TMV Interior.

Figure 21-3. Cabling System on Right Exterior of the Vehicle.

Figure 21-4. Transducer Storage Systems on the Left Exterior of the TMV.

plays) would be located to the left of the operator on the forward wall. Most data-handling equipment would be situated in an equipment rack located on the forward wall to the left of the densitometer panels. The L-shaped counter area was designed to provide adequate operator knee clearance, and for placement of clipboards, manuals, and other documents.

Figure 21-5. Roll-out Tray for Storage of a Heavy Transducer.

The layout also provided good lines of sight to the treatment. When seated at the work station, the operator would not block the customers view of the treatment through the window. The design team documented all of these dimensions and recommendations, with top and side view drawings, and clearances were checked by assembling a simple mockup of the van and through measurements made with design templates.

Displays. A long counter was provided for the customer's display, the plotter, and the printer. Each monitor was shock-mounted on a swivel stand and placed as close to the rear of the van as possible to minimize the amount of surrounding window area. The plotter and the printer were mounted on the counter for easy access by the customer and the computer operator. Support for remote, dedicated summary displays was also provided, especially for the service supervisor who might be located in another vehicle on the job site.

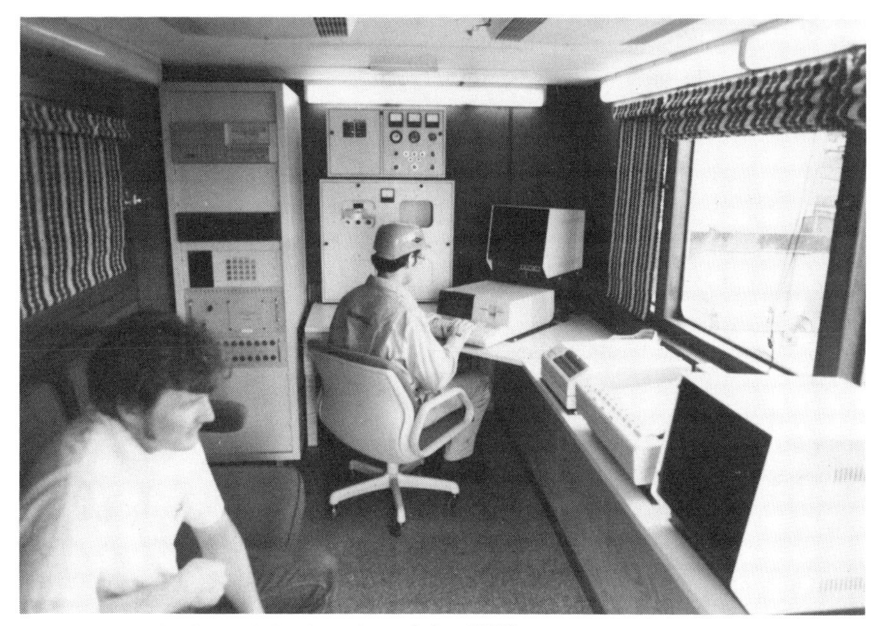

Figure 21-6. A View of the Interior of the TMV.

Equipment racks. An equipment rack, located to the left of the operator's station, contained a data acquisition and control unit as well as other components. The location provided for easy access during operation as well as during maintenance.

The universal power supply (UPS) power supply was located in the rear of the vehicle due to its size and the manner in which it was to be used. Although important, the UPS- associated controls would be operated only at the beginning and end of a job. This provided sufficient justification for locating this unit in a secondary equipment rack.

Doors and windows. Recommendations for door and door window dimensions were provided. Recommended window sizes and placement were based on the requirement to people leaving the vehicle a clear view of individuals walking up the steps or on the platform outside the door. For safety reasons, the door opened outward.

Large main windows were tinted, and roll-down blinds were provided to reduce the light levels within the vehicle and to block the sun when it was low on the horizon. The bottom of the window was no more than 10 in above the counter top, which allowed an observer seated on the bench to see the ground a reasonable distance from the vehicle.

General work space and environment. Two ceiling-mounted air conditioners were provided. Ceiling heights were 6 ft 5 in from the floor to optimize head clearance. This clearance accommodated over 90% of all operators wearing boots and hard hats. Fluorescent light panels, located over the operator's work station, the printer, and the plotter, provided interior lighting. A dark washable material, which minimized reflections in the displays, covered the walls.

Recommended Display Formats and Dialogs

A general approach to the design of the operating system was developed in which all tasks required of the operator would be specified on the basis of treatment phases. The basic phases of fracturing service were identified, as were the activities required of the TMV operator during each phase. The originally identified nine phases, were later reduced to four phases:

- Treatment setup
- Treatment run
- Post job analysis
- Treatment data transfer

The most important way in which the computer would aid the operator is the operations-oriented menu for the management of all TMV data and systems. menu was structured along the lines of treatment phases. The largest data input task occurred during the setup phase, during which the operator described the well and the service to be performed.

The design team incorporated many human factors principals into the data entry procedures, including minimizing of keystrokes, escape/cancel/correction capabilities, cursor pre-positioning, highlighting, field length specification, and fixed and predictable sequences of questions. Menus were also configured with a "map" to refer to current location within the menu structure. Function key labels were also provided on-screen for each menu page. Extensive recommendations regarding the selection and design of color displays addressed convergence and distortion, borders, use of saturated colors, contrast and brightness, color selection and viewing environment, line width, symbology, and overall format and layout of graphics.

The computer was programmed to print a transducer hook-up list. The operator used the list to string cables and attach each cable to its designated transducer, all according to instructions provided by the computer. As a result, this feature eased transducer deployment and minimized the frequency of cabling errors.

Evaluated TMV in Operation

Two reviews of operational equipment were conducted, the first was a checklist review prior to actual operation in the field. The second review was conducted during the course of a TMV-monitored fracturing treatment in Oklahoma. Overall, the system performed well and required one-half the time to cable and configure than an older, research data acquisition vehicle, which monitored the same job. Deployment of the 13 production vehicles was conducted with very few problems, and the TMV introduction was called the "most trouble-free of any vehicle ever developed for use by the field" within the company.

Developed Operator Training Material

At the request of the client, the human factors team developed the training materials for the first group of TMV operators. The team developed training objectives, prepared an extensive training course manual, and determined responsibility for each of the major segments.

Developed TMV Operator's Manual

The team also developed an operations-oriented manual for operators of the Treatment Monitor Vehicle. The human factors team members were in an ideal position to write the manual, based on their experience developing the interface and determining operational features. All phases of TMV operation were documented and defined with task lists. The task list for each phase of operation was structured into checklist items, each of which was to be performed during the course of a job phase. All menus and options were illustrated and defined, and supporting sections, such as trouble shooting were prepared. The 180-page manual was produced and included as an integral part of the TMV.

CONCLUSIONS

Human factors methods and data can have a major impact on systems when they are incorporated at the earliest phase of system development. Such was the case in this project, in which human factors involvement was used from conceptualization to deployment. The result was a system that met all original performance objectives, and was well received by its operators and other users.

The human factors professionals in this project were Steven Casey and James Gutmann, Anacapa Sciences, Inc. The Treatment Monitor Vehicle was developed and deployed by the Dowell Division of Dow Chemical.

Steven M. Casey is President of Ergonomic Systems Design, Inc. He received a PhD and an MS in Psychology/Human Factors from North Carolina State University and a BA in Psychology from the University of California. Dr. Casey is a member of the Human Factors Society and has served twice on the Executive Council. He is also a member of the Society of Industrial and Organizational Psychologists and the Society of Engineering Psychologists. His primary professional interests are in human factors engineering and the design of heavy equipment.

Section VI

Macro-ergonomics

22

Macro-ergonomics: A New Tool For the Ergonomist

David C. Alexander, PE

Macro-ergonomics is a relatively new concept. It views work from the system and organizational perspective instead of from the task and sub-task level. As a result, macro-ergonomics examines the organizational structure, the interaction of people inside the organization, and the motivational aspects of work.

Macro-ergonomics has advantages in its own right and can also provide a foundation for implementing a traditional ergonomics program. One key to the successful use of ergonomics is to have widespread knowledge of ergonomics by man people in the organization including operators, supervisors, managers, engineers, and health and safety personnel. Macro-ergonomics ensures that all people in the organization are provided a high quality of work life and provides the organization with the traditional aspects of performance—productivity, quality, and safety.

ERGONOMICS VS. MACRO-ERGONOMICS

Ergonomics

Ergonomics strives to improve worker performance by ensuring that people are compatible with their workplaces and job task requirements. That goal is met through two general procedures, one reactive and the other proactive.

The reactive approach to ergonomics is the most traditional and the most widely used. This approach identifies an existing problem and corrects it. As noted in Figure 22-1, a problem is any mismatch between the person performing a task and the requirements of the task. For example, when a person has to lift a package, the weight of the package must be within the strength capabilities of the lifter. Problems are identified in a variety of ways, but typically surface with injury/illness reports or through ergonomic inspections. Once a problem is identified, a solution is developed and implemented.

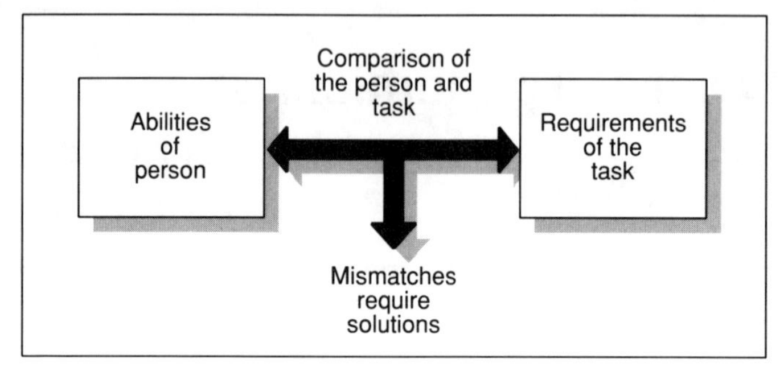

Figure 22-1. Ergonomics: The Person-task Match/Mismatch.

The solutions do not have to be sophisticated to be effective. While automation of problem tasks is often evaluated, other solutions can be "low tech" (changing tools or equipment) or "no tech" (altering workplaces, methods changes). As outlined in Table 22 -1, using a simple problem solving approach allows the ergonomist to define the problem and then create ways of dealing with it.

Table 22-1. A Problem Solving Process for Ergonomic Problems.

Step 1:	Identify potential problems
Step 2:	Compare task with human capabilities, and/or verify that the problem really exists
Step 3:	Develop alternative solutions, including both engineering and administrative solutions
Step 4:	Select the best solution
Step 5:	Implement that solution
Step 6:	Follow-up

The proactive approach to ergonomics requires the designer to anticipate human/task mismatches, and then to design so these problems are prevented. The proactive approach uses engineers for the identification of problems, and requires that they be skilled at anticipating human performance problems in the early stages of design. Again, a wide range of solutions remain available, and do not necessarily require technical sophistication or burdensome costs.

It is best to utilize both the reactive and proactive approaches to ergonomics, because each approach has strengths that the other does not. Ergonomics seems to work best when it is widespread at both the problem identification and solution development stages. This is one reason that macro-ergonomics is so well suited as a means to resolve traditional ergonomics concerns.

Macro-ergonomics

Macro-ergonomics is also the study of work, but it examines work and work systems from a broader viewpoint than that traditionally used. Some important issues in macro-ergonomics are the organizational structure, the interaction of people inside the organization, and the motivational aspects of work. While ergonomics has a focus at the task level, macro-ergonomics will focus at the task level, macro-ergonomics will focus at the organizational level. The building blocks for ergonomics are motions and micromotions; the building blocks for macro-ergonomics are jobs, teams, and groups of workers. Therefore, the ergonomics of tasks and jobs falls within the overall sphere of macro-ergonomics of tasks and jobs falls within the overall sphere of macro-ergonomics, as noted in Figure 22 -2. The idea is to look at the entire organization and not just its component parts, to examine both the forest and the trees.

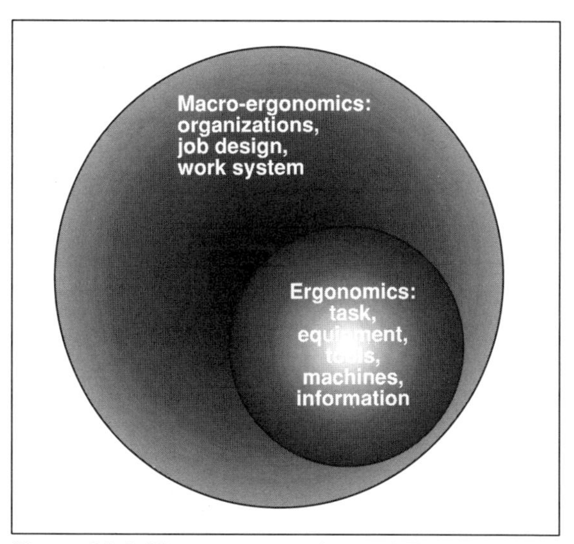

Figure 22-2. Macroergonomics and Ergonomics.

Hal Hendrick discussed macro-ergonomics and placed things in a useful perspective in *Macroergonomics: A Concept Whose Time Has Come.*[1] He describes ergonomics (human factors) as a series of three generations. The first generation was the

> focus on the design of specific jobs, work groups, and related human-machine interfaces, including controls, displays, work space arrangements, and work environments. Much of the human factors research was focused on anthropometry, perceptual, and other human physical characteristics in terms of their implications for hardware design. This development and application of human-machine interface technology continues to be an extremely important aspect of the science and practice of human factors.

Hendrick described the second generation which "increasingly emphasized the cognitive nature of work as reflected in systems design." And he continues, "Thus, the human factors design emphasis entered into the development and application of *user-system interface technology*."

The third generation is described:

> One result of the progressively increasing automation of factory and office systems has been a concurrent increase in our awareness of impact of technology on organizational systems. We have begun to realize that it is entirely possible to do an outstanding job of *microergonomically* designing a system's components, modules, and subsystems and yet fail to reach relevant system-effectiveness goals because of inattention to the *macroergonomic* design of the system. Although applied within a systems framework most of the work of the first two generations was focused on the design of specific jobs, work groups, and related human-machine interfaces. In short, these human factors efforts have been at the individual, team, or, at best, subsystem level, and thus have represented applications at the microergonomic level. The emerging third generation of human factors is focused at the macroergonomic or overall organization-machine system level and concentrates on the development and application of *organization-machine interface technology*. Macroergonomics begins with an assessment of the organization from the top down, using a sociotechnical system approach to the organization and system design. The notion here is that one cannot effectively design specific atomistic components of a sociotechnical system without first making scientific decisions about the overall organization, including how it is to be managed. Questions must first be answered concerning the optimal degree of complexity, centralization, and formalization that should be ergonomically designed into the structure.

Ergonomics and macro-ergonomics are not in conflict, they are, in fact, compatible with each other. A comparison of these two concepts is shown in Table 22-2.

In industry, macro-ergonomics is also known as organizational design (OD) and is used in the design of the organizational structure and reporting relationships. It looks at the way in which people work with each other and how the design of the organization influences that working relationship. Thus an organization with a flat organizational structure and wide spans of control will have easy access from the top to the bottom of the organization, but relatively limited contact between the worker and their supervisor at the shop floor level. Only an individual assessment of each organization can determine the most appropriate organizational design in which all the design variables are weighted appropriately. Some of the dimensions to be considered are found in Table 22 -3.

Table 22-2. A Comparison of Ergonomics and Macro-ergonomics.

	Ergonomics	Macro-ergonomics
Level of Detail	Micro	Macro
Unit of Work	Task, sub-task	Group, division
Goal	Optimize worker	Optimize work system
Focus	Details	Broad Overview
Measurement Tools	Generally physical measures such as length, force, lumens, decibels, time	Generally organizational and/or subjective measures such as number of people, span of control, attitudes, morale
Research History as Part of Ergonomics Field	20-40 years	3-5 years
Application History	10-20 years	1-2 years
Application Skills	Anatomy, physiology, perceptual psychology, industrial engineering	Organizational behavior, organizational psychology

Macro-ergonomics, like traditional ergonomics, seeks to involve the employee in business operations. There are several means of doing so and they are discussed in the following section.

Employee Involvement Programs

Employee involvement programs proliferated during the 1980s as more and more organizations adopted this tool as a way to enhance productivity, improve quality, and help control costs. These programs have been the most common form of organizational development taking on many forms, from brief "crew toolbox meetings" to full scale participative management programs and autonomous work teams.

Quality Circles. Quality circles are the most well-known of the employee involvement programs. Quality circles are small groups of employees with similar jobs and/or responsibilities who meet regularly one hour each week to solve problems related to their work area. They are often supported by facilitators who help the circle use the problem solving process and group dynamics, and assist in obtaining resources for the circle. The format for the quality circle concept is relatively constant. Each circle identifies problem areas, gathers relevant data, performs analyses, develops and justifies solutions, and implements the results. A structured approach to problem solving is often used as shown in Table 22-4.

Quality circles were originally conceived to resolve quality related problems, but their scope has expanded to include other aspects of performance

Table 22-3. Dimensions for Organizational Design.

Formalization	Amount of written documentation
Specialization	Degree to which organizational tasks are subdivided
Standardization	Extent to which similar work activities are subdivided
Hierarchy of Authority	Reporting relationships and span of control
Centralization	the level in the organization where decisions are made
Complexity	The number of activities or subsystems within the organization
Professionalism	The level of formal education
Personnel Configuration	The deployment of people for various jobs, as measured by the administrative ratio, clerical ratio, professional ratio, and direct/indirect ratio
Size	The number of people in the organization
Organizational Technology	The actions, knowledge, and techniques used to change inputs to outputs
Environment	Elements outside the boundary of the organization that affect the organization, including competitors, government, and customers
Goals	The unique purpose of the organization

Source: Organization Theory and Design, Richard L. Daft, West Publishing Company, St. Paul, MN, 1986, pps. 16-18

including productivity, cost control, and working conditions. The results of quality circles are work improvements, financial returns, and a feeling of ownership toward the organization.

Table 22-4. Problem Solving Process Used by Quality Circles.

Step 1:	Select a problem
Step 2:	Analyze the problem
Step 3:	Create solutions
Step 4:	Choose the best solution
Step 5:	Recommend solutions
Step 6:	Implement solutions
Step 7:	Follow-up

Crew toolbox meetings. A simplified form of employee involvement is the crew toolbox meeting. These meetings, as the title suggests, are held with employees gathered around their toolboxes. The agenda of these meetings is often safety or general business announcements. Meeting length, typically 10 to 15 minutes, is so short that the meetings are usually intended as a one-way transfer of information to the employee. While there is little time for the detailed problem solving found in quality circles, there is time for input from employees on problem areas. Thus the toolbox meeting may be good for identifying problem areas and for sharing information with the workforce. In practice, any problem solving that occurs is largely the result of work by individuals after the actual meeting.

One advantage of a toolbox meeting is that it is easy to convene and requires little planning to be successful. The toolbox meeting can be used prior to the start of a new or particularly hazardous task, thus ensuring that safety and sound work practices are on everyone's mind.

Work teams. Work teams are more complex forms of employee involvement than quality circles and toolbox meetings. Work teams consist of the crew members and their supervisor. The team makes work assignments based on available personnel, and when there are extra personnel, trains new members in skill development. General duties such as housekeeping, safety, training, and relief are shared by the crew members. Special assignments such as representation on department safety committees are rotated among crew members. Pay is based on skills known by the person rather than on the daily work assignment. Flexibility of work duties, an understanding of the entire work process, and the ability to perform any task are highly prized and thus pay the most. Workers with few skills are usually new to the organization and are paid the least. The pay/promotion sequence consists of developing the additional skills to perform all the team tasks. Even people with special skills (e.g., electricians) are encouraged to train others, at least in the basic skills of the trade.

Responsibility is a key component of work teams with the team being responsible for production and quality, as well as resolving problems and making things work. This is reflected in their pay—they are often given a salary rather than an hourly rate. They have input into hiring decisions, and may assess the progress of newer workers. The team or team members identify problem areas, develop appropriate solutions, and then implement the necessary changes. Ownership is a word often used to describe the feeling of being part of a work team.

Autonomous work groups. Autonomous work groups take the concept of a work team even further. These groups have full responsibility for all activities associated with production of goods and/or operation of a unit. While work teams often have a supervisor (titles such as coordinator or facilitator are often used), the autonomous work group is self-managed.

Once again, the spirit of owning the business is strong. Just as people will make modifications to tools or equipment that they own, the ability to change things often allows these groups to modify their workplaces and tools so that they are easier to use. With traditional work relationships, people are reluctant to make changes because "that's the way things are" and the result is poor user interfaces and low effectiveness.

A CASE STUDY

A large producer of synthetic fibers wanted to revitalize an old operation which had begun in the 1930s and had experienced many changes and upgrades over the years. Nonetheless, profitability was low, and morale was declining. A socio-technical approach was evaluated and decided to be used to improve the operation.

The Design of the Team System

A team system was designed to enhance communications within the work group and with line supervision. A team was composed of a supervisor and the employees under their direct supervision. This is based on the concept of using the supervisor as the link to the next layer of management. Each work team was given specific work assignments and control of the resources to accomplish their job. The teams were established based on normal organization charts. Responsibility and authority were delegated downward, and a feeling of ownership was created.

Each team was given time to assess its operations, identify operating problem areas, and develop solutions. While the team system was not designed to foster ergonomic projects, there were a number of ergonomic benefits as a result of the project. The feeling of ownership and the responsibility for operations resulted in the identification of many "ergonomic-type" problems that the team itself could solve.

Some ergonomic improvements. One such problem involved a clean-up operation in a mid-deck area where the ambient temperature was relatively high, especially during the summer months. The team recognized the problem, contacted appropriate safety and health resources, and developed administrative controls to avoid excessive heat stress.

Another problem area was in the remanufacturing of component parts where the final step was a buffing operation. A simple jig was developed to avoid unnecessary holding of small parts during buffing. This reduced a significant risk factor for cumulative trauma disorders, especially carpal tunnel syndrome.

In another area small parts required critical inspection before they could be installed as part of the manufacturing equipment. The parts were inspected under a microscope with very critical lighting. The team identified the need for changes to the lighting, and also found different eyepieces for use on the inspection microscope. Both changes enhanced the quality of the parts and reduced operator fatigue.

Job rotation was implemented on several tiresome jobs. Since the team members identified the need for job rotation themselves, it was implemented easier and faster than most rotation schemes. The operators understood the need for job rotation and developed a rotation schedule that was functional. The result was one administrative control that required little enforcement to maintain.

A second administrative control was also addressed. The use of safety glasses is never easy to enforce, yet one team decided which areas needed eye protection and then enforced this safety rule among themselves and with other workers. The peer relationship made enforcement much more effective and also ensured that safety glasses were worn even when supervision was absent from the work area.

When casters on carts are used in a fibers plant, the casters are sure to load up with excess fibers. The result is casters that are difficult to roll, even with significantly higher push and pull forces on the carts. Ensuring clean casters is typically an administrative nightmare—the results are poor and the frustration level is high. In this plant, however, the team understood the need for easily rolling carts and made an effort to clean the casters more frequently. Prior to the team system, eight carts were cleaned per day; after the system was installed, forty-five carts were being cleaned per day.

Reaches and clearances are always a problem with older production equipment and so was the case in this facility. Some of the older equipment needed to be cleaned in hard to reach areas. The idea of using a rake to reach the area was brought up in a team meeting. It provided an easy and functional way to reach the area and allowed the equipment to be kept cleaner and more functional. The use of a tool to extend the reach of people is an old trick of many ergonomist, yet it was a natural idea of the team.

Another ergonomist's trick is the use of larger wrenches when additional torque is needed. When disassembly of used production equipment proved difficult, the team found that a larger torque wrench helped reduce the difficulty of the task.

Mixing up inventory parts had always been a problem in one of the supply storerooms in the production area. Various alternatives had been tried over the years, but none was completely successful, and the problem was simply live with. The team was not satisfied with this level of performance however, and began to examine ways of reducing the error rate. Being more careful worked for a while, but the real solution was to color code the inventory parts by user area.

Repair of material handling equipment is not always done in a timely manner and the result is excessive force by the user with the possibility of injury. The teams developed a way of monitoring this equipment and obtaining needed repairs quickly. For example, a monorail developed a worn spot on the track and was becoming difficult to move, especially for female operators. The team identified this as serious problem and got maintenance to repair it quickly. In the past, similar problems were not reported or were not repaired quickly, and the result was material handling equipment that did not function well.

As parts became dirty, they were removed from the production equipment for cleaning. Since the parts were impregnated with a material which hardened and made later disassembly difficult, the parts were disassembled in the production area immediately after removal from the machines. The tools used for this operation were "off-the-shelf" tools, and were not really satisfactory for the job. The team worked to improve the tools for this disassembly operation.

Observations on Work Team Concept

Ergonomics seeks to identify those areas in which there is a mismatch between the person's abilities and the requirements of the job to be performed. Identification of these mismatches is made easier when the actual performer is involved in the identification of problem tasks. When the team members themselves are invited to list problems that they have with performance, many ergonomic problems will emerge. The use of Quality Circles and other types of employee involvement teams has shown that small groups of workers can enhance quality and productivity.

In addition to the ergonomic benefits described here, the use of the team system resulted in cost savings of over $9 million with an investment of less than $2 million in time and effort. The improved communications, faster responsiveness, and heightened moral were also substantial benefits resulting from the implementation of the team concept.

CONCLUSIONS

As employee involvement programs are being considered, they need to be examined from two viewpoints:

1. To what extent do they enhance the implementation of traditional ergonomics?
2. To what extent do they fulfill macro-ergonomics needs?

It is not difficult to find satisfactory examples of employee involvement programs in industry. These are one form of macro-ergonomics where the study of work is done at a "macro" or organizational level. There are other forms of organizational development that focus on improving effectiveness through the study of organizational make-up and structure. All of these fall under the umbrella now being called macro-ergonomics.

At the same time, the use of macro-ergonomic concepts is a suitable way of implementing traditional ergonomics within an organization. By asking workers themselves what types of problems they are experiencing many of the person/task mismatches will be uncovered. The workers are able to identify genuine concerns and allowing them to resolve these problems often results in cost-effective engineering solutions or in effectively implemented administrative controls.

Reference

1. Hendrick HW. Macroergonomics: a concept whose time has come, *Human Factors Society Bulletin*. February 1987;30(2):1.

23

Ergonomic Circles in Assembly Line Manufacturing

Susan M. Gleaves and James J. Mercurio

Lithonia Lighting, the nation's leading manufacturer of commercial industrial lighting equipment, employs approximately 5,200 workers at 12 different locations in the United States and Canada. In order to progress and not just to maintain their leading edge in today's highly competitive market, Lithonia's corporate safety department and senior management embarked on an exciting company-wide ergonomics program to address the industry's most pressing safety problem—cumulative trauma disorders (CTDs). This decision only came after a concentrated effort to reduce the incidence of carpal tunnel syndrome showed that many CTDs were prevalent in Lithonia's assembly operations, especially at the older facilities.

Lithonia's senior management was well aware of the difficulties of having safety, engineering, and medical professionals trying to understand the differences among people and modifying the work environment to accommodate these differences. Consequently, senior management decided to make this program a cooperative effort, and involve all employees, both management and labor, in some aspect of the program. At the heart of the program are working ergonomic circles, which have proven very effective.

PROGRAM OBJECTIVES

Lithonia Lighting's ergonomics program began with a commitment of support from upper management and the corporate safety department. A steering committee was formed and given specific responsibilities. Members of the steering committee developed an outline of the initial ergonomics program and presented it to senior management. The initial program plan included a mission statement, objectives, and the plan for implementing the program. Additionally, the program outline included who would be involved in the program and how the program would be managed.

The mission statement was simply stated as: "Management and labor's teamwork on identifying and developing ways of making work tasks easier, safer, and more productive." The mission statement included six objectives:

1. To reduce individual exposure to injury-prone human movements
2. To reduce the potential of wrist, shoulder, and back injuries
3. To reduce workers' compensation costs associated with CTDs
4. To reduce unnecessary and time consuming movements at specific work stations
5. To improve teamwork and communication between management and labor
6. To create a foundation for future operations

COMMITTEE APPROACH/RESPONSIBILITIES

In the past, companies have used committees and task teams to address safety, quality, and production concerns. In recent years, the use of committees has expanded, employing Japanese-style management strategies, which in part rely on the teamwork concept. Lithonia's approach relied heavily on the committee approach, and was designed to ensure input from every affected area.

Two teams were developed, a steering committee and an initial task team, which later expanded into the Ergonomic Circles. The following outlines the membership of each team:

Steering Committee - Team Members
Chairperson—Director of Manufacturing/Plant Manager
Co-Chairperson—Facility Safety Representative
Vice President of Operations
Director of Plant Engineering
Director of Safety and Loss Prevention
Director of Manufacturing Services

Ergonomic Circles - Area Task Team Members
Facilitator—Department Manager
Area Industrial Engineer
Facility Safety Representative
Area Supervisor/Foreman
Hourly-Role Models—One or Two Hourly Employees

Additional group members included other hourly employees, maintenance, quality control, design engineers, purchasing, or others as needed.

Responsibilities of the Steering Committee were two-fold. Initially, this committee provided direction on major projects the area Ergonomic Circles were pursuing. Meetings were held on an as-needed basis to review project recommendations, financial needs, and the implementation of reactive, corrective needs. Additionally, the Steering Committee met with senior management to formally present ergonomic projects and to review the overall status of the ergonomics program.

The area task teams, which are at the heart of the program, are the Ergonomics Circles. Team members identify ergonomics problems and suggest improvements. The team members decide jointly on which items receive the highest priority. Having made those decision, each member is assigned a project or portion of a project to investigate and report back to the team. The entire team, however, follows up on the status of ergonomics projects.

Ergonomic Circles at Work

The initial implementation of the ergonomic circle concept started with one department. Since involving hourly workers in key decisions, would affect production and quality, which was new to Lithonia, both management and labor cautiously participated in this initial attempt. Soon, the foundation of mutual benefit and trust was established, which greatly increased employee involvement and communications. The initial success prompted Lithonia to expand this concept to other departments, facilities, and other operating division. The initial ergonomics circle functional format became the blueprint for the company-wide ergonomics circle program. The format follows:

- The facilitator, usually the department manager, coordinates team meetings, documents all action items, records the status of each item, and is the person responsible for investigation. The facilitator reviews the minutes of each meeting and provides the overall direction for the Ergonomics Circle.
- The facilitator formally schedules weekly, one-hour long meetings, which all team members regularly attends.
- The team identifies problems in its area and as a team prioritizes them. Team members are assigned responsibility for investigating or resolving specific items. The team, after developing and implementing specific solutions, then reviews the effectiveness of those corrective actions.
- Time outside the weekly meeting is set aside for team members to individually work on their assignments. The time requirement varies greatly with the project.
- Monthly, or as needed, the area team and Steering Committee meet to jointly review the team's progress, review priorities, discuss additional assistance, and allocate financial resources.

The Steering Committee targeted the assembly department at one of Lithonia's largest fluorescent manufacturing facilities to begin the ergonomics program. One assembly line, with approximately 40 workers, which assembled a family of fixtures that best represented Lithonia's stock-type fixtures, was chosen for the initial ergonomics team.

The plant manager and department manager jointly decided which two hourly workers would participate in this program. The workers were chosen based on their willingness not only to communicate worker concerns but also to participate in resolving these concerns for the benefit of both the company and affected workers.

Once these individuals were chosen, the team facilitator scheduled the team's first meeting. An outline of the program was distributed to each team member. The facilitator reviewed in detail the mission of the ergonomic circle and the team's objectives, the Steering Committee and its responsibilities, and how the team, via a formal organizational structure, would achieve its mission.

The facilitator asked each member to survey the workers in their immediate area and develop a list of items, concerns, or needs the team could address. This was to be accomplished outside the team meeting. The team reviewed the list and then toured the area to visually determine whether concern was justified and the extent of the problem.

At that time, the team listed the most critical items on a problem-action tracking sheet, starting with the highest priority. The team initially kept the list to a workable load, showing completed items while tracking new items added to the list. For each item listed, a member was assigned the responsibility of investigating that specific concern. At following meetings, team members reported on each item until successful resolution of that problem or other appropriate action had been taken. This format was repeated with additional Ergonomics Circles on other lines and in other departments.

Lithonia's company newsletter spotlighted the progress of the area task teams' efforts in making ergonomic improvements in their work areas. This let workers throughout the plant share in the ergonomic success stories. As more ergonomics teams were formed, the sharing of successes became a regular occurrence.

On the first anniversary of the program, six ergonomics teams were honored at a formal luncheon. The team members, middle management, and senior management representing three divisions within Lithonia Lighting attended the luncheon. Each team member received a world class trophy for individual and team efforts.

Soon, the team members recognized that many small adjustments do improve the overall work environment. Obstacles, such as the time it takes to implement corrective actions and to develop several alternative solutions, are hurdles each team has had to overcome. Getting engineering, purchasing and quality control employees to understand and react to ergonomics needs were also obstacles to the initial team's success. With time, however, the effectiveness of each team became readily known as its solutions and preventive measures successfully eliminated some of the most pressing problems and reduced injury frequency and severity.

The team concept proved beneficial from the outset. Hourly workers would propose a change in work methods, while engineers would propose new tooling or design changes. The safety representative would present injury trends to the team or discuss a specific accident that would require the ergonomics team's input. Lithonia was quick to recognize the benefit gained by this combined labor-management effort.

Continued Ergonomic Circle meetings highlight the program's efforts. Since its inception, adjustments to the meeting schedules, team membership, frequency of meetings, and project load have been made to satisfy the needs of individual departments and to keep the program on a dynamic, proactive course.

ERGONOMIC IMPROVEMENTS

Ergonomic improvements as a result of the program have ranged from very simple, inexpensive workstation modifications to the purchase of mechanical systems designed to reduce fatigue, eliminate excessive motions, and increase worker acceptance of tasks. Some of the workplace modifications included:

- Hanging all air-tools from overhead framework, with a balancer and overhead air hose, thus reducing CTDs and trip hazards
- Providing small carts for parts, which allows easy movement to and from the assembly line, thus reducing bending and lifting hazards
- Providing tables for pallets of parts thus reducing lifting hazards
- Tilting tables for various tasks in several departments to effectively keep parts in the "neutral" or "safe" zone of every worker
- Adjusting work heights when stacking fixtures on pallets by scissor lift-tables
- Replacing old rubber casters with larger, solid phenolic wheels on many floats, carts, and dollies
- Installing a material handling device, which uses a vacuum system to aid in manual handling of heavy fixtures
- Using anti-vibration (impact) gloves at stations requiring bumping action
- Using fatigue mats at stations that did not previously have mats
- Increasing lighting at some work stations

Some of the engineering modifications included:

- Changing from flat tipped screws to pointed screws
- Increasing the hole tolerances on some fixtures to be assembled together as well as decreasing the screw specifications

- Providing more heat to assembly lines via overhead infrared heaters
- Providing anti-torque sleeves and grips that are used with specific air tools
- Reducing the force required to insert ballast wires into sockets
- Evaluating air shut off tools designed for use in light assembly type operations
- Increasing the use of the telescoping torque arm for in-line air screwdrivers

Many other improvements have been made by hourly workers, industrial engineers, and supervisors of Lithonia Lighting. One of the most dramatic changes was the ability of labor and management to pull together as a team, and together attempt to resolve some of the major problems facing the lighting industry today. Change at Lithonia Lighting has become synonymous with improvements to the quality of the work environment, a main objective of the ergonomics program at Lithonia.

ERGONOMIC TRAINING

As with any new program, Lithonia Lighting recognized that specific training needs must be addressed during the implementation phase of the new Ergonomics Circle program. Back injuries, specifically low back strains, and CTDs are among the lighting industry's major safety and medical concerns.

Lithonia Lighting embarked upon a five-part Back-Injury Prevention Program (BIPP), with the fifth session dealing solely with ergonomics and manual handling tasks. All hourly workers received this training as part of their monthly safety meetings. This training was implemented alongside the development of the ergonomic task teams.

A two-part Cumulative Trauma Disorder Training Program was also developed and implemented. Medical, safety, and engineering controls were reviewed and site-specific risks and control methods were presented. Management, supervisors, and some ergonomics teams have gone through this CTD training. Additionally, a CTD booklet was compiled and distributed as part of the CTD Training Program.

ERGONOMIC IMPACT AT LITHONIA LIGHTING

The impact of the ergonomics program at Lithonia Lighting manifests itself in several ways. The initial ergonomics team concept has encouraged many operations to involve the hourly labor force in other committees and decision making tasks teams. It has also created the foundation for future projects, such as World Class Manufacturing, which relies heavily upon employee involvement.

Process improvements play a major role in reducing ergonomic problems through automation and work station redesign. Ergonomics is a prime consideration for all purchases of new equipment and for major modifications of existing equipment. Office ergonomics, through the development of Video Display Terminals (VDT) Ergonomic Guidelines, has included office employees in the ergonomic revolution.

The prevention of problems has thus become a major focus of Lithonia's' ergonomic efforts, including such groups as design engineers, purchasing agents, and maintenance personnel. Ergonomic Circles, task teams, and teamwork all point to the future mentioned in Lithonia's ergonomic mission statement. Thus, the embodiment of ergonomics at Lithonia Lighting has fast become a way of life.

Susan M. Gleaves is a Consulting Engineer with Auburn Engineers, Inc. She has both a Bachelor and Master of Science in Industrial Engineering from Auburn University. Since joining Auburn Engineers, Susan has been a consultant for OSHA and has conducted ergonomics training for OSHA compliance officers, as well as providing ergonomic and safety consultations to a wide range of industrial clients. Susan's current interests are concentrated on occupational ergonomics, with a special emphasis on cumulative trauma disorders, workplace design, worker/equipment compatibility, and ergonomics/safety interactions.

Section VII

Maintainability

24

Maintainability Design

James W. Altman, PhD

The purpose of maintainability design is to maximize the availability of a product to the customer through scheduled or failure-initiated actions. Such design is reduced to a trivial level by decisions to take no action to prevent or to correct malfunctions during the product's operational life. Otherwise, maintainability design becomes important. Figure 24-1 suggests the general structure of maintainability design considerations.

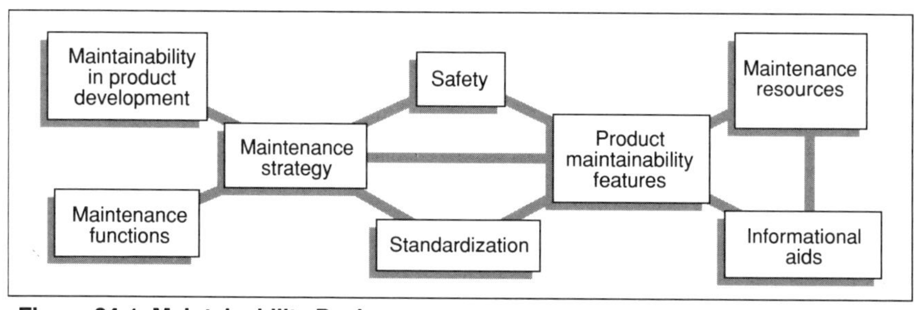

Figure 24-1. Maintainability Design

First, we look at the role of maintainability design in product development, distribution, and operation. Second, we suggest maintenance functions, which may need to be supported by design and development actions. Third, we identify major decisions in the development of a maintenance strategy. Fourth, we highlight safety considerations in maintenance. Fifth, we suggest relationships between product standardization and maintainability. Sixth, we describe design features impacting product maintainability. Seventh, we suggest support resources to complement product design characteristics in achieving maintainability. Eighth, we address issues in presenting information to maintenance personnel.

MAINTAINABILITY IN PRODUCT DEVELOPMENT

Maintainability supports a product's functional and performance objectives. Effective maintainability design considers the larger of product planning and engineering.[1,2] Figure 24-2 suggests five main developmental considerations for maintainability. Each is addressed in a separate section below.

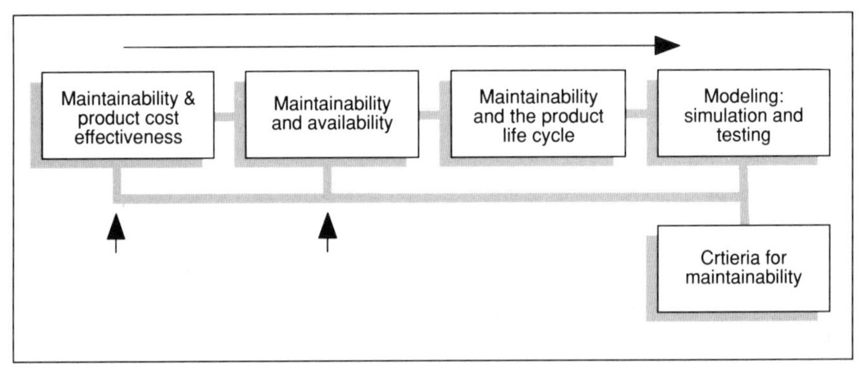

Figure 24-2. Product Development.

Product Cost Effectiveness

Quality of maintenance is important to producer long-term well being because, with the possible exception of quite inexpensive or robust products, effective maintenance tends to be an important factor in customer satisfaction.

Availability

Maintenance is detrimental to customer satisfaction when it is costly or disruptive. A product whose operational life meets or exceeds customer expectations, which requires no preventive maintenance, and is 100% reliable will be fully available. A robust and reliable product that never requires maintenance is clearly best. But, limits on achieving reliability are likely to require allowance for maintenance in the life of a complex and costly product.

Product Life Cycle

Maintainability design must be coordinated with product planning, design, development, marketing, distribution, and use. Figure 24-3 suggests what we consider to be an appropriate relationship between maintainability design and the rest of the product life cycle. A maintenance strategy should be generated as part of early planning. Specification and design of product maintainability features should be accomplished in interaction with the product architectural and component design. Maintainability design should be implemented and tested in parallel with other aspects of product development.

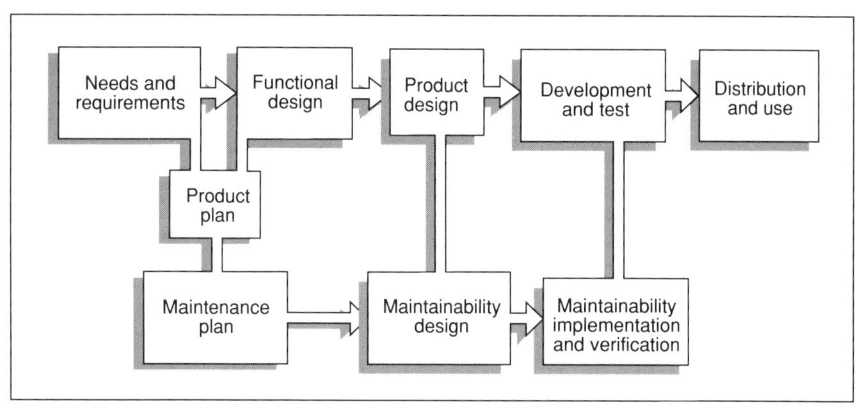

Figure 24-3. Maintainability Design in the Product Life Cycle

Maintenance Modeling, Simulation, and Testing

Maintenance modeling, simulation, and testing provide a basis for judging which alternative design best meets maintainability criteria. A *maintenance model* is any representation of maintenance of the product. *Simulation* involves the manipulation or exercising of a model to generate estimates of maintenance performance. *Testing* involves performance of maintenance tasks on the product or a close prototype thereof. Together, these techniques can play an important role in assuring effective product maintainability.

Criteria For Maintainability

An effective maintainability design minimizes:

- Down time for maintenance activities or as a result of maintenance-induced failures
- User or technician time to accomplish maintenance tasks
- Logistics requirements for parts, backup units, tools, and personnel
- Equipment damage resulting from attempted maintenance activities or failure to accomplish needed maintenance
- Personnel injury resulting from maintenance actions or from failure to accomplish needed maintenance
- Costs resulting from maintainability features and maintenance costs during operational use of the product.

MAINTENANCE FUNCTIONS

Each maintainability design feature is justified by its ability to support needed maintenance functions. These functions can serve one or more of three principal maintenance purposes.

- Preventive: To help to keep the product in good condition
- Corrective: To return a malfunctioning product to usable condition
- Recycle and overhaul: To return a product to its approximate original condition.

The following seven maintenance functions address preventive, corrective, and recycling purposes:

- Operation: Activates the product or support equipment toward the achievement of maintenance objectives.
- Handling: Supports maintenance by assembling, installing, disassembling, removing, connecting, disconnecting, loading, unloading, opening, closing, unpacking, packing, or transporting.
- Service: Enhances fault-free performance by changing, charging, fueling, cleaning, decontaminating, heating, cooling, lubricating, or replenishing.
- Verification: Assures proper condition of the product or identifies symptoms of improper condition through physical inspection or functional checking.
- Analysis: Interprets fault information to localize the source of trouble (also commonly known as troubleshooting).
- Fault correction: Restores the product to proper working condition through
  - *Adjustment* (alignment, calibration, tightening, loosening, or tuning) to compensate for slippage, wear or drift
  - *Repair* (joining and mending, sealing, unbinding, separating, shaping, unsorting and rerouting)
  - *Replacement* to trade defective components for new

- Management and support of maintenance: Includes communication, transportation, procurement, supply, planning, scheduling, maintenance performance monitoring, quality control, safety and health monitoring, damage analysis, cost minimization and budgeting, personnel management, and management of maintenance records and data.

MAINTENANCE STRATEGY

Developing a maintenance strategy for a new product follows the general sequence shown in Figure 24-4. In addition to the development cycle and maintenance functions discussed above, three additional kinds of design considerations are involved in developing a comprehensive maintenance strategy: degree of maintenance support for the product; level of automation planned for maintenance activities; and characteristics of the maintenance personnel subsystem. Each of these decisions is discussed further below. Once a comprehensive strategy is developed, it will guide each of the areas of maintainability design listed in the right hand box of Figure 24-4 and discussed in the final five main sections of this chapter.

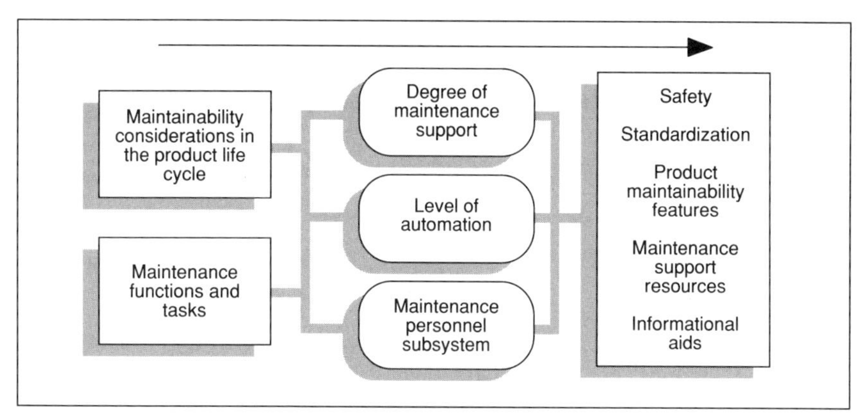

Figure 24-4. Development of Maintenance Strategy.

Degree of Maintenance

The degree of maintenance needed for a given new product will depend upon decisions of four principal kinds:

1. The extent to which product reliability will be sought and assumed to establish or preclude a need for maintenance
2. The segmentation of the product into modules for fault detection and correction
3. The provisions made for repair or discard of the various modules of the product
4. The provisioning of backup modules

Reliability and maintainability. Maintainability design decisions relate to reliability considerations in two ways: (1) preventive maintenance can enhance product reliability, and (2) corrective maintenance involving minimum product down time can also minimize the effects of unreliability on product availability.

Maintainability design should start with functions for which there is no reasonable alternative and then proceed to tradeoffs based on reliability, maintainability, and cost considerations.

Module definition. A module is a unit of the product separable for purposes of failure prevention or correction. Relevant optimizing criteria[2] include:

- Minimum cost, including considerations of whether modules are to be discarded or repaired
- Optimum size, where size may be defined by physical volume, weight, physical components, circuits, functional elements, and unit cost
- Minimum time for fault location and correction
- Minimum burden from inter-module connectors
- Maximum standardization, with as many of the modules being interchangeable as possible
- Homogeneity of life expectancy for components within modules so that the remaining life of components within a discarded module will be minimal
- Minimum supply burden by restraining the number and variety of units required to be stocked, especially at remote field locations
- Minimum field impact from corrections of design deficiencies and field enchancements of the product

Repair or discard decisions. Repair-discard decisions impact the maintenance organization, management, and support structure. The pivotal time for deciding throwaway points is during design since optimum design of a unit can depend heavily upon whether or not it is intended to be repairable. Total life cycle cost should be a major consideration in repair-discard decisions. Important cost factors are:

- Acquisition—including design, development, production, distribution, and installation
- Maintenance personnel—including training and job pay
- Maintenance Material—including replacement parts and modules, test equipment, tools, supplies, facilities, working space, storage space, utilities, and documents
- Transportation—including shipping, handling, and associated communications

Some of the factors favoring a discard policy include:

- Reliability gain by avoiding the adverse effects of technicians accidentally injuring equipment.
- Manufacturing costs by facilitating miniaturization or permanent sealing
- Maintenance simplification by avoiding the requirements for within-module skills

- Modification simplification—by enabling plug-in of improved units rather than field rework
- Costs savings from simplified design; from the economies of volume production; and from reductions in training, in numbers of maintenance technicians, in repair facilities, and in performance aids

Some of the factors favoring a repair policy include:

- Elimination of connectors for replacement modules, with resulting size and weight reductions
- Reduced storage space through the use of smaller piece parts rather than module spares
- Use of current repair facilities rather than their abandonment to a throw-away process
- Greater capability in the field to react to emergency situations and to innovate
- Cost savings by using relatively inexpensive piece parts rather than more costly replacement modules

Provisioning. The provisioning of spares is the part of any effective maintenance program. The essential elements in the formulation of a provisioning policy include:

- The delineation of replaceable modules for the product
- Estimation of reliability for each module
- Estimation of the number of the product that will be operational
- Estimation of the times required for each phase of spares production and supply

Given such information, optimum strategies for keeping all essential modules in supply are amenable to modeling and simulation techniques. Management of a provisioning program is usually best supported by a computer-based inventory system. Powerful commercial off-the-shelf (COTS) software packages are available.

Level of Automation

Major relationships of maintenance functions to automation considerations are as follows:

- Operation of the prime product will be determined principally by performance rather than by maintenance objectives. This is especially true of test equipment, whether plug-in or built-in.
- Handling of the product is generally manual in the field, but can be highly

automated in production lines modified to support product recycle or in major maintenance depots.

- Service automation is almost always technically feasible, with cost considerations usually being the determining factor.
- Verification through functional test is typically economical to automate, with improved reliability and greatly improved speed over manual procedures. Physical inspection of visible surfaces is generally best left to manual procedures, although inspection in hazardous environments or for microscopic or sub-surface damage usually requires special sensors and mediated display of sensed data.
- Analysis of malfunction data for diagnostic and troubleshooting purposes by automated means requires a profound capture of product design expertise, but is readily feasible in almost all instances. The equipment and systems software to support such analysis have become economical, but the costs of knowledge capture and representation in applications programs can be substantial.
- Fault correction in the field will usually (with the frequent exception of adjustment) be most cost-effective if it is essentially manual. Robotic replacement is feasible and economical on a production line modified for product recycle.
- Management and support can usually benefit from a computer-based system to capture, store, retrieve, process, and report maintenance transaction data.

Maintenance Personnel Subsystem

A maintenance strategy for a given product or family of products must include a consideration of the human resources that will be available to implement it. These considerations fall into two major areas.

- Maintenance roles and levels—defining how required maintenance functions and tasks will be allocated to echelons of maintenance (from end user of the product to the factory), to organizational units at a given echelon, and to positions within organizations.
- Personnel requirements—defining staffing, skills and knowledge, recruitment, selection, training and assignment of personnel to accomplish product maintenance.

Safety

Safety should be a paramount concern in all aspects of maintainability design. Maintenance is frequently the most strenuous and dangerous facet of a product's life cycle. Important occupational health and safety considerations for maintenance include:

- Locating
  - accesses (entrance ways, approach areas, etc.) to minimize the likelihood of accidental contact with electricity, extreme temperature, toxic substances, and moving parts.
  - internal controls away from dangerous voltages;
  - overhead-mounted components such as pipes so that technicians will not bump them; and
  - bolts, nuts, lugs or screws in recesses or protected configurations to avoid protrusion into the technician's path.

- Providing
  - catches to hold open heavy access covers that might fall shut and cause damage or injury;
  - holding devices for components that may slide or tilt and injure technicians;
  - molded grips to fit the fingers and prevent side pressure when units must be carried frequently or for long periods of time;
  - safety interlocks on accesses leading to equipment with high voltages;
  - automatic activation of an internal light where accesses are located over dangerous mechanical components;
  - self-locking fasteners to prevent danger to the technician or to equipment in the event of failure or loss;
  - vibration-resistant fasteners, locking mechanisms or supporting structures for equipment that is stowed overhead, to prevent its falling and injuring personnel;
  - mechanical safety-locking devices on hydraulic jacks to prevent accidental lowering of the load in the event of hydraulic system failure;
  - handles on crane hooks to preclude injury to fingers when technicians must hold or guide the hooks during lifting operations;
  - independent brakes on hoists—a solenoid-released, spring-operated brake and a mechanical load brake, each capable of holding 1-1/2 times the full rated load on the hoist;
  - total enclosure or screen guards for all moving parts on service equipment;
  - wrapping on sling cables to prevent injury to personnel and damage to the equipment being lifted;
  - locking devices to preclude turntable boom movement when cranes are being transported;
  - ladders of nonconductive, splinter-proof light-weight construction with chemical-resistant materials, non-slip rungs and bases, lashing devices for portable ladders, and carrier rails for fixed ladders.

- Protecting edges by rounding, smoothing or covering with protective material. Use rounded edges and corners on the exterior surface of units and mountings. Avoid sharp or thin edges on grips.

- Insulating tools intended for use around electrical currents.
- Labeling accesses with warnings against hazards that might be encountered at the point of access.

Standardization

Standardization is an important and pervasive factor in product design, production, and life cycle costs. One of the areas in which standardization has important effects is in maintenance. The following illustrate some of the ways in which standardization impacts maintenance:

- A minimum variety of fastener sizes, threads, and head types will minimize supply problems, restrict the number of tools required of maintenance technicians, and the likelihood that use of an improper tool will result in bungled maintenance.
- Standardization is the key to cost-effective supply, especially in the field.
- Interchangeability of components within and across equipments reduces supply problems—but physical interchangeability is to be avoided where components are *not* functionally interchangeable.
- Consistency of nomenclature and labeling is an important factor in helping the technician to follow maintenance instructions and procedures.

Product Maintainability Features

Maintainability features include accesses; controls and displays; connectors, couplings, and lines; fasteners; handles; labels; mounting and positioning; test and work points; and unitization.

Accesses. Major features in the design of accesses are their location, size, shape, and covers. Accesses should be located where they are convenient for the maintenance technician without interfering with operations, free of obstructions, and oriented directly in front of internal work points so the technician will not be required to reach to the side or behind obstructing components.

Access sizes should easily accommodate 95% of the technician population. This is accomplished for one-handed tasks as follows, for:

- Insertion of an empty hand with the fingers extended through a square hole. The size of the access should be at least 3-1/2" if the hand is bare, and 5-1/2" if it is covered with a mitten.
- Insertion of an empty hand held flat. The size of the access should be at least 4-1/2" on the major axis of the hand is bare and 6-1/2" if the hand is covered with a mitten. On the minor axis, 2-1/4" are required if the hand is bare and 5" are required if the hand is covered with a mitten.
- Insertion of an empty hand held as a fist. The size of the access should be at least 5-1/8" by 4-1/4" if the hand is bare and 7" by 6" if the hand is covered with a mitten.
- Reaching through an access to the depth of the elbow. The size of the

access should be at least 4-1/2" by 4" for a technician dressed in summer clothing and 7" by 6" for a technician dressed in heavy winter clothing.
- Reaching through an access to depth of the shoulder. The size of the access should be at least 5" by 5" for a technician dressed in summer clothing and 7" by 6" for a technician dressed in heavy winter clothing.
- Insertion of a hand holding a 1" diameter handle screwdriver through a square hole. The size of the access should beat least 3-3/4" if the hand is bare and 5-3/4" if the hand is covered with a mitten.
- Insertion of a hand holding a component. The size of the access should be at least the dimension of the component in the axis of insertion plus 1-3/4" if the hand is bare or plus 3-3/4" if the hand is covered in a mitten.

Accommodation to technician hand size is accomplished for two-handed tasks as follows, for:

- Reaching to a depth of 6" to 25 ". An access height of 4-1/2" is required for a technician dressed in summer clothing and 6-1/2" for a technician dressed in heavy winter clothing. An access width is required which is at least 75% of the depth of reach for a technician in summer clothing. An additional 2" allowance is required for a technician in heavy winter clothing.
- Reaching to a depth greater than 25". An access height of 13" is required for a technician dressed in summer clothing and 6-1/2" for a technician dressed in heavy winter clothing. An access width of 20" is required for a technician in summer clothing and 32" for a technician in heavy winter clothing.
- Insertion of component grasped by the sides to a depth up to 8". The size of the access should be at least the width of component + 4-1/2" for bare hands and 7-1/4" for hands in heavy mittens.
- Insertion of component grasped by front handles with no clearance for hands necessary. The size of the access should be the width of the component + 1".

Accommodation to technician body size is accomplished in design for technician passage as follows, space for:

- Entry of the technician's head through an access should be 8" by 8" if bare and 10" by 10" if in a winter hood.
- Entry of the technician's shoulders through an access should have a width of at least 20" in summer clothing and 32" in heavy winter clothing.
- Entry space for the thickness of a body passing through an access should be 13" for a technician in summer clothing and 21" in heavy winter clothing.
- The space for a technician crawling through an access should be 31" high

by 20" wide for a technician in summer clothing and 34" high by 32" wide in heavy winter clothing.
- Entry through an access in a kneeling position with back erect should be 64.5" high by 20" wide for a technician in summer clothing and 67" high by 32" wide in heavy winter clothing.
- The space allowed for two standing technicians passing through an access abreast should be 36" wide if they are in summer clothing and 52" wide if they are in heavy winter clothing.

An additional consideration in designing accesses is to make them large enough to permit simultaneous physical and viewing access or to provide dual windows to serve both physical and viewing access needs.

The shape of accesses should facilitate operation of their covers. Access openings should be designed in a shape that will facilitate passage of the required items for maintenance tasks. The openings need not necessarily be round, square, or rectangular. Using asymmetric shapes, keys, or mounting holes will safeguard against the mismounting of access plates that must be attached in a singular position (e.g., to prevent malfunction of units attached on the back of the plate).

Covers should serve reliability objectives with minimum interference to maintenance tasks. Use coverless openings wherever they will not degrade reliability or esthetic standards. Provide windows for visual inspections where coverless openings will not meet reliability criteria. Use quick-opening (e.g., hinged and sliding) covers where coverless openings or windows are not appropriate.

Controls and displays. The same ergonomic principles as apply to product use apply to maintenance controls and displays. However, maintenance is especially likely to demand control operation with tools, in conjunction with operator displays from unusual control locations, and in conjunction with special test instruments.

Connectors, couplings, and lines. A large portion of maintenance time is spent with the disconnection and connection of lines and components. Substantial savings can be achieved if the number of connectors and lines is kept to a minimum, standardized, arranged for ease of maintenance, simple and reliable to operate, and easily identified. Lines such as wires, cables, tubing and piping tend to be obtrusive in maintenance tasks. They should be located so they do not have to be removed unnecessarily in order to perform maintenance on other equipment units.

Fasteners. The number of fasteners should be kept to the minimum commensurate with product integrity and safety. Operation of fasteners tends to consume a substantial portion of maintenance time. In fact, the number of fasteners involved with a maintenance task is one of the best predictors of maintenance time. Detailed recommendations for fastener design can be found in Folley and Altman (1956a, b), Altman et al (1961), and Crawford and Altman

(1972) and in general human engineering design guides.

Handles. Some of the considerations in design of handles to facilitate maintenance are as follows:

- Provide handles on units which are either too bulky to be carried easily, or are too small to grasp and hold.
- Recessed grips should be provided near the back of heavy units to facilitate handling where protruding handles would interfere with an effective operational configuration.
- Hinged handles should have a stop to lock them in their extended position.
- Recessed finger lifts may be used on units under 25 pounds where conventional handles are not acceptable.
- Quick-connect handles should be used where permanent handles would interfere with operational configurations.
- Physical characteristics of handles should include:
    - a grip width of at least 4-1/2" for a bare hand and 6-1/2" for a hand covered in a heavy mitten;
    - a depth of at least 2" for a bare hand and 3" for a hand covered in a heavy mitten;
    - a diameter of at least 1/4" on 10 to 15 pound units, 1/2" on 15 to 20 pound units, and 3/4" on units over 20 pounds (but not larger than 1-3/4");
    - four grip spaces for units over 40 pounds (requiring lifting by two technicians), either by providing four standard size grips or two grips large enough to accommodate two hands each.

- Location of the handles should provide
    - a center of gravity of the unit below the grip while the unit is being lifted and carried;
    - no contact of the handle or hand with controls, other units or danger;
    - a comfortable position for removal and replacement of the unit as well as carrying; and
    - protective support by using handles as maintenance stands.

Labels. Labels should follow general ergonomic principles and should facilitate maintenance tasks, for example:

- Align labels on access plates with the structures to which they are to be attached, to facilitate proper positioning of the plate.
- List items (such as test and service points) to be reached through the access and service equipment.
- Uniquely designate each component with which the technician must interact so each can be identified in job instructions.
- Indicate position for insertion of components and connectors through small accesses. Use matching stripes, cots, or arrows on the cabinet and on

the component to be inserted. Where space permits, Provide a drawing of proper alignment positions.
- Support easy distinction between test and service points (e.g., using color and/or shape coding).
- Provide luminescent marking for test points, service points, and other components that must be used under conditions of low illumination.
- Warn of danger by clearly marking potentially dangerous items and areas. A standard configuration for warnings should be used. An inverted triangle is preferred.[6]

Mounting and positioning. The attachment and placement of components should provide accessibility for maintenance with consideration of use factors such as frequency of removal; interference with other components; size and weight of the component to be mounted; space requirements for access, removal, and replacement of the component; and maintenance in place on the installed component.

Consideration should be given to methods of mounting such as:

- Roll out, slide out, or hinged racks for easy checking of components that must be pulled out of their normal installed position (provide limit stops and braces to stabilize components in the open position).
- Twist or push-to-lock mounts for small components (provide a locking screw to assure against vibration or a detent where vibration is not a threat).
- Shock mounts for instances where vibration might interfere with operations such as fine calibration.
- Spring clamps to mount tubing, pipes, or wiring which may require frequent removal and replacement.
- Friction lugs to secure inaccessible sides of components.
- Easily mated mounting studs (especially where visibility is poor).
- Guide pins or their equivalent on units for proper alignment during mounting; dimpled holes, pointed bolts, or other provisions to facilitate the starting of bolts through openings; and a minimum number of fasteners consistent with stress considerations.

Consideration should be given to component location and factors such as:

- A removal path along a straight or slightly curved line and free of blockage by other components.
- Symmetrical brackets to assure correct positioning.
- Isolation of delicate components from areas where frequent or heavy maintenance is likely to be required.
- Clearance for tools in the area around mounting fasteners and between all units for ease of removal and replacement.

- Continued activation of maintenance controls and displays even though panels and other components may be removed for maintenance.
- A working level of 34" to 60" from the floor for standing technicians and 17" to 43" for the seated technician.

Test and work points. Test points provide access to information about equipment status. Service points provide access for lubricating, filling, draining, charging, and similar service functions. Maintainability design considerations for test and service points include location, connection, and design.

Location of test and work points should:

- Be compatible with the frequency and sequence of maintenance tasks. Highly accessible test points should be provided for system verification.
- Require minimum disassembly and removal of other equipment.
- Provide access at the surface of equipment when it is assembled or installed.
- Provide a comfortable work height (4-1/2 ft to 5-1/2 ft for the standing technician).
- Permit access through a single opening to all of the clustered test and service points used in a given routine.
- Isolate service and test points from dangerous areas.

Connection of test and service equipment should be provided for by:

- Accesses which facilitate entry for internally located test and service points.
- Holding devices for test probes and service nozzles where the technician may require the use of both hands for adjusting controls, etc.
- Direct insertion or quick connectors (unless high pressure demands a threaded connection).
- Receptacle keys to prevent access to points where pressures or potential fluid loss may be a problem unless the proper device is fully inserted.

Design of test and service points should provide:

- Access sufficient to avoid a need to disassemble equipment for verification, fault location, or servicing (except for major recycle and overhaul).
- Needle probe testing of hermetically sealed units.
- Built-in high pressure indicators to avoid some of the dangers involved in temporary high pressure connections.
- Special "ground" points if good grounding points are not available and connection to ground should be made during tests of a given unit.
- Guides for test probes when test points are internally located.

Unitization. Unitization decisions under the section above on degree of maintenance will have a profound impact on product unitization as it affects maintenance. Some additional considerations include:

- Unit size and weight which are compatible with the available tools and handling equipment—neither too small or compact nor too large or heavy for handling by the technician team. Where possible, make units small and light enough (45 pounds or less for lifting heights of five feet or less) for one technician to handle and carry. Provide for two-technician lift where weight does not exceed 90 pounds. Units weighing more than 90 pounds should have provision for mechanical or power lift. Microelectronics have been found to reduce maintenance burden but require special tools for optimum performance.
- Independent verification and adjustment, if required, of the individual unit should be possible.
- Firmware tends to be extremely small, complex, and expensive. Layout, mounting provisions, and tools should support damage free removal and replacement.
- Software (the code) does not involve maintenance per se, but the medium (e.g., electromagnetic, optical) may degrade and require replacement or renewal of the code on the same medium. Provision must be made in maintenance design for the maintenance of backup copies of code and for restoration or replacement of damaged code. "Bugs" sometimes do not show up on software until after it becomes operational. Maintenance plans must provide for removal of defective versions of software and replacement with corrected versions. Also, the installation of software updates and enhancements may be considered part of the system maintenance process.

MAINTENANCE SUPPORT RESOURCES

Ergonomic considerations may have an important role in selecting and designing maintenance support resources. These resources might include work environments and facilities, vehicles, dollies and other special transporters, platforms, ramps, ladders, walkways, power and hand tools, test equipment, and bench mock-ups. Guiding principles may include a concern for minimizing the requirements for such equipment, simplicity, freedom from error-inducing characteristics, and compatibility with all relevant maintenance especially those of limited use; simplicity and freedom from error inducing characteristics; and compatibility with and availability for all relevant maintenance procedures.

Successful maintainability design must incorporate consideration of the effects of the working environment on technician performance. The environment in which maintenance is to be accomplished can have a profound effect on the efficiency. Environmental factors must be considered in design to assure rapid, accurate, and safe maintenance. Work environment, facility factors, support equipment, and tool considerations are presented in detail in general human engineering handbooks and in the context of maintainability design.[3, 4, 5, 7, 8]

INFORMATIONAL AIDS

Informational aids are descriptions of procedures in a form suitable for use by maintenance technicians. Below are presented some considerations in relating information ergonomics to maintenance tasks.

Standard Procedures

Reliable performance of every maintenance procedure requires informational support in the form of knowledge remembered by the performer, cues inherent to the maintenance environment, or special informational job performance aids. Principal types of informational aids are:

- Photographs and drawings to identify equipment items, illustrate important features, and show significant spatial or functional relations.
- Instructions that guide the user or technician through procedures. These are best provided as a series of steps, each described briefly, either as an outline series or as flow diagrams with instructions for each step as a block. Warnings and safety instructions should be clearly highlighted and differentiated from steps having neither damage nor safety implications. In computer-based products it is effective to provide instructions (help files) stored on the computer. Audio instructions are especially useful as support to early or infrequent task performances.
- Keys, which cross reference between aids and equipment, should be maintained.

Analysis and Fault Localization

Periodic verification of proper product functioning should be a standard procedure. Analysis of results from such routine testing requires only a determination of whether or not all indications are nominal. Once a pattern of malfunction is detected, however, the process of diagnosing possible causes and selecting the most efficient sequence of localizing checks can become substantially more complex—with some evidence[9] that the delineation of a consistent set of possible faults is more difficult than the selection of a near-optimum test sequence to localize within the set of possible faults.

For more than three decades there has been a movement away from requiring technicians to diagnose faults from an understanding of the product design. This emphasis on diagnostic rather than design information for the technician has had a favorable impact on training requirements and task performance.[10, 11, 12]

It has also been found that analysis and fault localization (troubleshooting) performance can be facilitated by computer-aided problem structuring and that such computer aid can have positive transfer effects on similar problems for which no computer aiding is available.[13] Computer-based problem structuring can help technicians to perform more like an ideal Bayesian processor—that is, making full use of available information. Applications of artificial intelligence (AI) techniques to fault diagnosis and localization have been found to be feasible and effective in fields such as marine systems,[14] large telecommunications networks,[15] and aviation.[16] Promising results[17] have also been obtained in applying AI techniques in the development of versatile general-purpose computer-based maintenance aids. Information technology has reached the point where serious consideration must be given to providing expert computer-based informational aid to maintenance tasks for all major new products.

References

1. Altman JW. *Some Procedures in Design for Maintainability*. Wright-Patterson Air Force Base, Ohio: Aerospace Medical Research Laboratories; 1962. Air Force Systems Command Publication MRL-TDR-62-9.

2. Goldman AS and Slattery TB. *Maintainability: A Major Element of System Effectiveness*. New York: John Wiley & Sons, Inc.; 1964.

3. Folley JD, Jr and Altman JW. *Guide to Design of Electronic Equipment for Maintainability*. Wright-Patterson Air Force Base, Ohio: Aero Medical Laboratory; 1956b. Air Research and Development Command WADC Technical Report 56-218.

4. Altman JW, Marchese AC, and Marchiando BW. *Guide to Design of Mechanical Equipment for Maintainability*. Wright-Patterson Air Force Base, Ohio: Aerospace Medical Research Laboratories, Air Force Systems Command; 1961. ASD Technical Report 61-181.

5. Crawford BM and Altman JW. Designing for maintainability. In: Van Cott HP and Kinkade RG, ed. *Human Engineering Guide to Equipment Design*. Washington, DC: American Institutes for Research; 1972:585-631.

6. Riley MW, Cochran DJ, and Ballard JL. An investigation of preferred shapes for warning labels. *Human Factors*. 1982; 24:737-742.

7. Folley JD, Jr and Altman JW. Designing electronic equipment for maintainability. *Machine Design*. June-December 1956a.

8. Johnson SL. Evaluation of powered screwdriver design characteristics. *Human Factors*. 1988;30:61-69.

9. Toms M and Patrick J. Some components of fault-finding. *Human Factors*. 1987;29:587-597.

10. Altman JW, Folley JD Jr, Wilkinson FR, and Brinda J, Jr. *A Study to Determine the Feasibility of Fully Routine Troubleshooting Procedures*. Pittsburgh: American Institutes for Research; 1959.

11. Elliott TK and Joyce RP. An experimental evaluation of a method for simplifying electronic maintenance. *Human Factors*. 1971;13:217-277.

12. Shriver EL and Trexler RC. *A Description and Analytic Discussion of Ten New Concepts for Electronics Maintenance*. Alexandria, VA: Human Resources Research Office; 1986. HumRRO Technical Report 66-23.

13. Morris NM and Rouse WB. Review and evaluation of empirical research in troubleshooting. *Human Factors*. 1985; 27:503-530.

14. Govindaraj T and Su YD. A model of fault diagnosis performance of expert marine engineers. *Int J Man-Machine Studies*. 1988;29:1-20.

15. Lee NS. DM^2: an algorithm for diagnostic reasoning that combines analytic models and experiential knowledge. *Int J Man-Machine Studies*. 1988;28:643-670.

16. Smith PJ, Giffin WC, Rockwell TH, and Thomas M. Modeling fault diagnosis as the activation and use of a frame system. *Human Factors*. 1986;28:703-716.

17. Shapiro SC, Srihari SN, Taie MR, Geller J, and Campbell SS. *The Versatile Maintenance Expert System (VMES) Research Project*. New York: Rome Air Development Center; 1988. RADC-TR-88-11, Vol II.

James W. Altman is President of Synectics Corporation. While there, he has been involved in the development of evaluative techniques for graduate training in social work, for nutrition education and for adult education through mass media. He has also been engaged in work with human factors in environmental protection, defense systems and with the development and application of technology to facilitate human interaction with information systems. Dr. Altman obtained his PhD in Personnel Psychology from the University of Pittsburgh.

25

Design of Mining Equipment For Maintainability

E.J. Kirk Conway and Richard L. Unger

Underground coal mining equipment is used in one of the most demanding operating environments known. As a result, the cost of maintaining this equipment is very high. In addition, the number of maintenance-related injuries remains persistently high despite efforts to reduce the number of accidents.

BACKGROUND

Thirty-five years ago underground coal mining equipment consisted of relatively simple but rugged machines powered by electric motors and hydraulics. These machines were used to cut, dig, load, and transport coal from the mine face to the surface. The machines were maintained by the mine maintenance personnel armed with a basic knowledge of hydraulics, electricity, and mechanical design. These maintainers were expected to repair all of the equipment at the mine site using simple hand tools.

Over the years the basic mining machine has been transformed into powerful, complex mining systems. To boost productivity, the horsepower and size of the original machines have been substantially increased, and machines were designed to perform multiple functions. To increase throughput, continuous miners, long and short wall systems, and continuous haulage were introduced. Also, numerous safety features and environmental control systems have been added to the machine to protect the miners' health and improve safety.

With few exceptions, however, little improvement in the basic design of equipment for maintainability has been made, and in many cases, equipment maintainability has sharply decreased. Many of the design changes were achieved by simply modifying existing machine design. On certain mining machines, sharp reductions in maintainability were experienced as a result of added-on safety and environmental systems. Even with all of the design changes, the maintainer is still expected to service and repair these ever more complex machines in an operational environment that provides little in the way of new maintenance tools, procedures, automatic test equipment, or other technology-based maintenance aids.

One major area of concern is the escalating cost of mining equipment maintenance. Underground equipment maintenance typically accounts for 25% to over 35% of the total mine operating costs. These costs have continued to increase over the last 15 years despite efforts to contain them. Mine operators have attempted to gain control of these steadily increasing costs through:

1. Optimizing scheduled maintenance operations
2. Reducing maintenance staff
3. Reducing and better controlling spare parts inventories
4. Contracting for maintenance support
5. Deferring non-essential maintenance

In addition to dollar costs, equipment maintenance has traditionally accounted for one-third of all lost-time injuries in underground mines. This injury rate persists despite concerted efforts by mine management to minimize accidents, Mine Safety and Health Administration (MSHA) to enforce health and safety rules, and the Bureau of Mines to conduct safety research.

Over the years, little attention has been focused on the design of the mining machine itself with respect to maintenance costs.[1] The cost of maintaining a machine is, after all, a direct function of:

- The maintenance frequency and failure interval of the machine and major components
- The time and labor required to complete unscheduled maintenance actions
- The time and labor required to complete routine maintenance tasks

A review of current mining equipment design suggests that considerable cost savings could be achieved with relatively simple design improvements.[1,2] For example, by relocating difficult to access, but frequently replaced, hydraulic valves and hoses on certain roof bolters, this one-hour plus removal and replacement (R/R) task is reduced to a five-minute operation. Numerous other maintenance savings could be realized with minor design improvements on new or existing equipment.

In addition to reduced task completion times, improved maintenance safety could be realized. Improved component accessibility and increased ease of using R/R parts reduces the maintainers' risk of injury. These design improvements also contribute directly to equipment productivity.

Equipment users and manufacturers recognize the need for improved design for maintainability.[1] Many mines, in fact, are currently modifying or rebuilding equipment to facilitate maintenance. Several mining equipment Original Equipment Manufacturers (OEM) are actively seeking ways to improve the maintainability design of their equipment. Two of these OEMs have impressive research efforts directed at incorporating new maintenance technology into their designs. At least one mining OEM is exploring innovative new technologies to enhance the overall maintenance process.

DESIGN INDUCED MAINTAINABILITY PROBLEMS

The US Bureau of Mines researchers analyzed underground coal mining equipment with respect to design for maintenance and maintenance personnel safety. A maintainability design review and human factors analysis of equipment was completed at nine operational coal mines.[1] Mining machines in large and small mines operating in high and low seam coal were surveyed, including conventional, continuous, and long wall operations. Shuttle cars, scoops, roof bolters, continuous miners, long wall equipment, undercut machines, face drills, utility vehicles, and personnel carriers were also reviewed. The survey identified several design limitations which directly impacted maintenance time, cost and personnel safety.

The primary design limitation was inaccessibility to components. In many cases the access openings were too small (Figure 25-1), the components were crowded into compartments, hoses and power cables overlayed or were run in the frame or chassis where maintainers could not reach them, and mounting bolts or connectors were hard to reach (Figure 25-2). As a result, maintainers often had to remove unaffected parts to get at the failed or suspended components or disassemble the machine to locate fasteners and mechanical interfaces (Figures 25-3 and 25-4). The crowded design also subjected components to impact damage and increased maintenance burden (Figure 25-5).

The mining machines design also made routine maintenance more cumbersome. Component crowding made visual and physical inspection very hard. Maintainers could not quickly R/R leaking hydraulic hoses, water lines, and failed hydraulic valves. Routine lubrication also took more time than necessary if the design was more conducive to maintenance.

Another limitation was the design's fault isolation capability. Consequently, maintainers had difficulty determining the precise cause and location of a failure. Part of this was due to limited or no designed-in fault diagnostic capabilities or effective failure indices.

A fourth problem was lack of resources, particularly of the proper tools.

Maintainers often had to "jerry-rig" tools, handle 100 lb to 1000 lb components, and substitute brute human strength to move components.

Figure 25-1. Inadequate Access Opening Size.

COST OF MINING EQUIPMENT MAINTENANCE

Reliable maintenance cost data are currently unavailable across the underground coal industry. Several industry estimates are available, but they vary substantially from source to source. Actual maintenance cost summaries were obtained from several participating mines—these figures are not available for publication, however.

Informal data gathered during this research project found that equipment maintenance costs range from 20% to over 35% of total mine operating costs.[1] Actual values varied based on the size and type of mine, mining technology employed, management attitude towards maintenance, and other factors.

These figures are substantiated by other published research. A British study,[3] for example, found that maintenance costs account for 30% plus of the total operating costs in underground British coal mines. Similar unofficial estimates have been obtained from two major American coal mine operators.

Figure 25-2. Poor Equipment Layout–Protective Covers Must be Removed for Routine Servicing of Hoses.

Figure 25-3. Inadequate Access Room–Components Under Motor Cannot be Reached.

Figure 25-4. Component Installed in Interior Cavity.

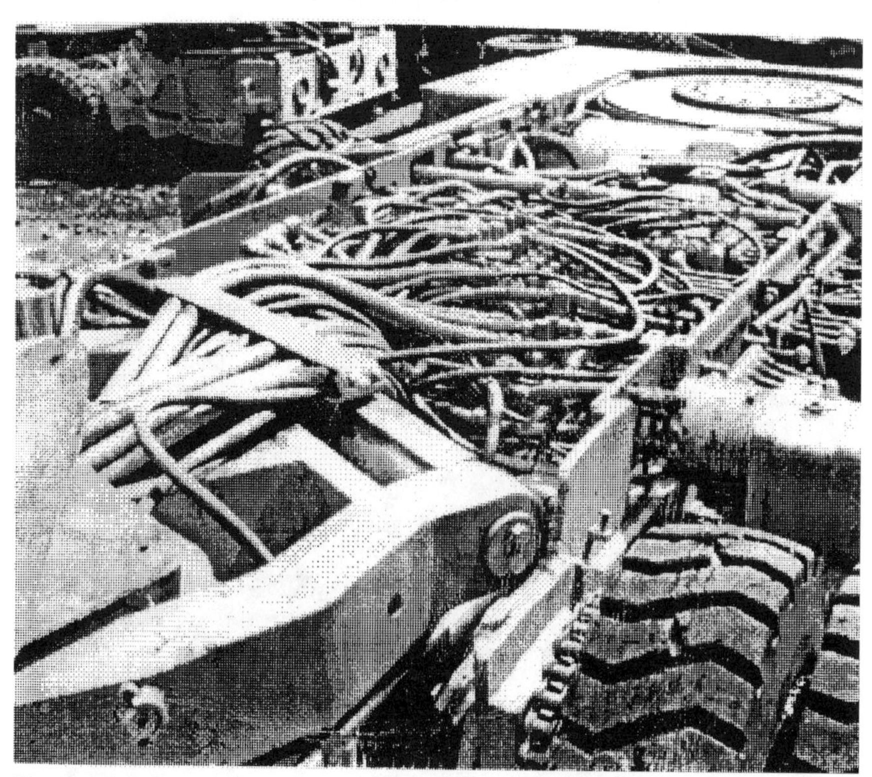

Figure 25-5. Crowded, Unorganized Equipment Compartment.

Factors Contributing To Maintenance Costs

A review of mine maintenance operations[1] suggested that management's attitude towards maintenance and the skill of maintenance management personnel contribute to equipment maintenance costs. Management attitudes range from "when it breaks–fix it" to strong top management support for professionally planned and implemented preventive maintenance programs. These programs are geared to reducing unscheduled equipment down time and to controlling maintenance costs. The skills required to organize and manage an effective mine maintenance program differ from the skills required to perform "hands on" maintenance of mining equipment. Poor maintenance management contributes to increased costs. Other factors that contribute to equipment maintenance costs include:

- Maintainers Training and Experience - Poor maintenance skills result from inadequate training; lack of job performance aids; manuals, and guides; and the complexity of maintenance tasks.
- Maintenance Environment - To maintain a continuous miner in a 36-in coal seam is an entirely different task than to maintain one in a well equipped 6 ft high underground repair shop. Maintenance can be completed more expeditiously in the latter environment, but it is not always possible.
- Age Of Equipment - Older equipment tends to be smaller and inherently simpler in design. As a result, older machines are somewhat simpler to maintain. Newer equipment tends to be larger, more complex and overlaid with numerous add-on systems and components, making accessibility and the basic maintenance process more difficult.
- Maintenance Errors - Reliable data are not available, but most maintenance personnel interviewed concede that maintenance errors contribute substantially to overall maintenance costs.[1] Removal and replacing non-failed items, trouble-shooting one system too long, not replacing suspected components during a previous maintenance opportunity, failing to install or repair a component correctly, failing to test a component prior to reassembly, and related errors account for an estimated 10% to 25% of all maintenance time.
- Design Of Equipment Itself - Certain makes and models of mining equipment are designed to facilitate maintenance and repair, while the basic design of other models hinder maintenance actions.
- Regulatory Compliance - Safety and environmental control devices required for regulatory compliance adds to the complexity and increases maintenance costs.

MAINTENANCE SAFETY COSTS

Maintenance operations account for a significant percentage of all coal mining accidents and injuries. Mine Safety And Health Administration accident statistics for 1984 suggest that maintenance-related injuries account for 33% of all lost time accidents.[4] These accidents impact mine operating costs in the form of lost productivity, increased benefits costs, and increased insurance rates.

Many injury accidents can be directly traced to equipment design.[1, 2, 5] Inadequate accessibility, lack of means to lift and maneuver heavy components, inability to visually observe the maintenance task being performed, inadequate maintenance safeguards, and other design-induced problems account for a significant percentage of maintenance accidents. Improved accessibility, enhanced component-machine interface, and simplified maintenance procedures could have a positive impact on these statistics. Improved maintenance safety will reduce maintenance as well as overall operating costs.

COST OF DESIGN FOR MAINTAINABILITY

The value or worth of any machine resides in its ability to generate a return on investment. If a machine has an initial cost of "Y" dollars, it must produce "Y +" dollars of coal to have a positive worth or value. On this assumption, it is possible to illustrate the cost savings derived from improved design for maintainability using simple economic models.

Many economic models can be used to compute the worth (W) of equipment. For this discussion, a simplified model will suffice. Figure 25-6 presents an overview of this model. (Readers interested in a more comprehensive treatment are referred to references 6 - 9.)

The following model suggests that the worth (W) of a piece of mining equipment can be defined as:

$$W = I + C + M - P$$

Where:

I = The initial purchase price of the machine
C = Cost per hour to operate the machine
M = Maintenance costs per hour of operation
P = Production value per hour of operation

The initial purchase price (I) of the piece of equipment is fixed or inelastic. It is set at the time of purchase. The price is simply amortized per hour over the useful life of the machine. Of course, the more hours of production, the lower the amortized cost per hour.

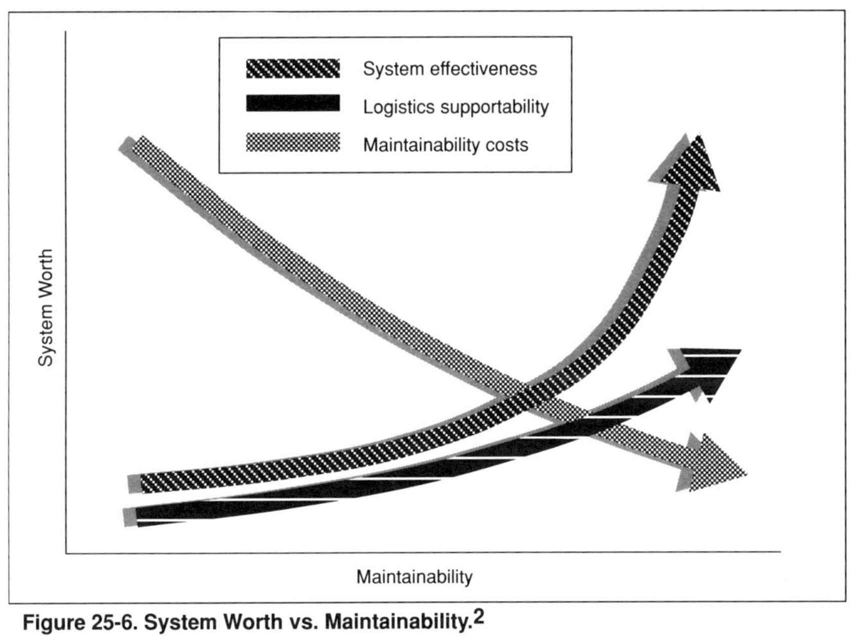

Figure 25-6. System Worth vs. Maintainability.[2]

The Cost Per Hour (C) to operate the machine is relatively fixed or inelastic and is composed of the following cost elements:

- Labor costs for the machine operator(s), support personnel, and immediate production supervision.
- General Overhead Costs which include insurance, utilities, royalties, brokerage, and related costs
- Cost of mining supplies and materials
- Other management and administrative costs

The cost to maintain (M) consists of the following cost elements, some of which are fixed and some of which are relatively elastic:

- Labor costs for maintenance personnel
- Cost of spares, replacement parts, and supplies
- Loss of production during maintenance
- Cost per hour of idled machine operators
- Other maintenance related costs

The costs of replacement parts and maintenance supplies are also relatively inelastic. Certain savings can be realized with careful buying. The cost of labor and other overhead items, on the other hand, are a function of the duration of repair time for unscheduled corrective maintenance actions.

A reduction in repair time for downed equipment contributes positively to the overall worth equation by increasing the time available for production. Thus, decreased time-to-repair not only reduces direct maintenance costs, but increases the production per hour thereby offsetting other costs. If we look at the maintenance process again, we observe many points at which time can be saved through improved design for maintenance (Figure 25-7). Several of these points include:

- Prediction of pending failures to facilitate PM scheduling
- Decreased fault isolation time
- Reduced component access time
- Decreased inspection and diagnosis time
- Diminished component removal/replacement time
- Reduced test/alignment time

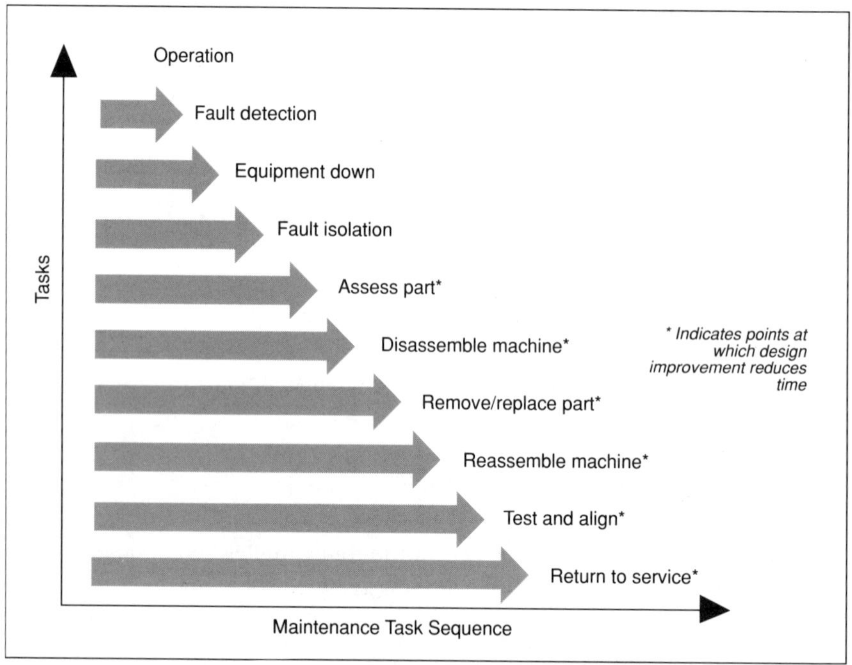

Figure 25-7. Sample Maintenance Task Sequence.

A review of underground maintenance task completion times at two large mining operation revealed that the time required to change hydraulic hoses on miners and shuttle cars ranged from 15 min to over 3 hrs. The estimated average time for a failed hydraulic hose R/R was over 35 min. Examination of these machines revealed that the time differences were directly linked to accessibility of the hose connectors. In several cases, two or more non-failed components had to be removed in order to access a failed hose connection.

By relocating several components or rerouting hoses, maintainers could directly access over 90% of all hydraulic line connections on the surveyed machines. This would have reduced the average hydraulic line R/R time to well under 15 min per replacement.

If a maintainability design standard for a new or rebuilt machine specified that all hydraulic hoses had to be R/R in less than 15 min, the average repair time for this task could be reduced 50%. Similar performance criteria could be developed for other maintenance tasks. The result would be significant reductions in all maintenance task completion times.

Evidence from other civilian and military research efforts suggest that preventive maintenance and corrective maintenance task time reductions of from 40% up to 70% can be achieved with planned maintainability design efforts.[8, 9]

PRODUCTIVITY

Productivity represents the other side of the maintainability issue. Productivity is a function of the machine producing coal, and is directly impacted by the speed and ease with which the mining machine can be repaired and returned to service. The more rapidly a machine can be returned to production, the more productive it will be.

Productivity is expressed in terms of the units (of coal) produced by a machine per unit of time. The greater the number of hours the machine is available to produce coal, the more productive it is going to be. For example, suppose that a continuous miner has a rated production capacity of 100 tons per hour. Further, suppose that the same miner requires an average of one hour of preventative maintenance (PM) per shift and one hour of corrective maintenance (CM) per shift. Assume that the mine operates the equipment during one production shift per day for 300 days per year. Hence:

300 hr PM + 300 hr CM x 2 shifts = 1200 hours per year

If the CM and PM time could be reduced by 50%, this would result in the following increase in productivity:

1200 CM + PM hrs/year /0.5 = 600 hrs per year savings
600 hrs x 100 ton = 60,000 ton/year/machine increase

If the mine were operating eight mining machines, this 60,000 tons per machine increase per year would be the equivalent of adding another miner with no additional increase in cost.

60,000 per year x 8 miners = 480,000 ton annual increase

Actual analysis of the design of three different continuous mining machines during this project suggested that productivity improvements exceeding the above example could be achieved with relatively simple redesign efforts.

EQUIPMENT DESIGN AND MAINTENANCE SAFETY

A summary of the underground coal maintenance related accident statistics for 1981 is presented in Table 25-1. A majority of the maintenance injuries involve strains/sprains, low back injuries, and crushing injuries. These injuries typically occur during R/R of components weighing from 35 lb to over 1000 pounds.[10]

In many instances two or more men with crow bars, 4x4's, or other makeshift tools must manhandle the component to remove it from the mining machine or to lift it into place so that it can be secured. In most cases, no provisions have been made during component-machine interface design to provide for mechanical assist in the R/R process.[1, 10] A review of mining equipment design suggests that in many cases, this designed-in assistance could be readily achieved. For example, if guide pins to hold components while they are being bolted or unbolted were added, personnel exposure to the types of injuries identified in Table 25-1 would be minimized and the R/R process itself would be expedited.

One of the objectives of maintainability engineering is to minimize the need to manhandle components into place. With proper design and engineering, all components should be provided with mechanical means to interface with the machine itself. With optimized maintenance design, it is reasonable to assume a substantial reduction in maintenance-related accidents.

HUMAN ERROR AND DESIGN FOR MAINTENANCE

Human error is a problem that must be addressed in design as well as during operation and maintenance of complex equipment.[5, 10-13] Errors occur in operating mining machines, performing maintenance tasks, or in making management decisions. Fortunately, most human errors result in limited negative consequences, for example, lost time and production waste. In many cases, the error ends up costing time or money. Unfortunately, in a smaller percentage of cases, people are injured or killed and equipment destroyed.

Operationally Induced Errors

What does human error have to do with mining equipment maintainability? In an interesting review of the subject, researchers report that a significant percentage of all operational equipment failures are human error induced.[12, 13] In fact, human error accounts for 50% to 70% of all electronics failures, 60% to 70% of all aircraft and missile failures, and 20% to 30% of all mechanical failures. Many of these are operator-induced errors resulting in machine damage or

Table 25-1. Maintenance Related Injuries (4).

		Mine Maintenance		Machine Maintenance	
Code	Type of Accident	Number	% Total	Number	% Total
1	Stationary object	185	5.6%	272	8.9%
2	Moving object	2		6	
3	Concussion	--		1	
4	Falling object	611	18.4%	511	16.8%
5	Flying object	62	1.9%	62	2.0%
6	Rolling object	62	1.9%	18	
7	--	--		--	
8	Struck by, NEC	231	6.9%	265	8.7%
9	--	--		--	
10	--	--		--	
17	Fall - walk away	8		7	
18	Fall on object	3		5	
20	Caught - meshing object	--		2	
21	Caught - move/stationary	169	5.1%	190	6.3%
22	Caught - 2 moving objects	6		9	
23	Caught - collapse	2		--	
24	Caught - NEC	261	7.9%	292	9.6%
25	Rub, abrade	3		1	
26	Bodily reaction, NEC	2		2	
27	Over-exert: lifting	1,132	34.1%	793	26.1%
28	Over-exert: pull/push	78	2.3%	147	4.8%
29	Over-exert: welding	112	3.4%	13	
30	Over-exert: NEC	360	10.8%	373	12.3%
33	Contact hot object	3		26	
36	Inhale noxious fumes	1		8	
38	Absorb noxious fluid	26		25	
39	Flash burns - electrical	--		2	
42	NEC	1		2	
43	Insufficient data	2		6	
	Total Within Category	3,322		3,038	

prolonged downtime. Maintenance requirements could be reduced by designing out these types of errors. Other errors are made by maintenance personnel while performing maintenance tasks.[14]

Maintenance Induced Error Rates

Rasmusen and Rouse[14] also report that 20% to 25% of all failures are directly traceable to maintenance errors. A separate study found 25% of all maintenance problems to be human error induced during maintenance operations.[12] Another study reports human error rates for specific types of maintenance tasks. These data, summarized in Table 25-2, were derived from an earlier study.[14] The values are indicative of the error rates found in many industrial and military settings.

It has been estimated that the average human reliability value in adjusting or aligning tasks is .0987.[14] This value suggests that out of every 1000 attempts to adjust a component, 13 errors can be expected. Many of these errors could be eliminated through improved design of the component-machine interface. Although not directly applicable to underground mining operations, the above error rates are suggestive of the types, frequencies, and sources of human errors in maintenance. Similar error rate patterns can be expected in mine maintenance operations.

Table 25-2. Representative Maintenance Task Error Rates.[14]

Action	Object	Error Description	Error Rate
Observe	Chart	Inappropriate switch action	1,128
Read	Gauge	Incorrectly read	5,000
Read	Instruction	Procedural error	64,500
Connect	Hose	Improperly connected	4,700
Torque	Fluid lines	Incorrectly torqued	104
Tighten	Nuts & bolts	Not tightened	4,800
Install	Nuts & bolts	Not installed	600
Install	O-rings	Improperly installed	66,700
Solder	Connection	Improper solder joint	6,460
Assemble	Connector	Bent pins	1,500
Assemble	Connector	Missing part	1,000
Close	Valve	Not closed properly	1,800
Adjust	Linkage	Improperly adjusted	16,700
Install	Orifice	Incorrect size installed	5,000
Machine	Valve	Wrong size drill & tapped	2,083

*Error rates per million operations

Errors In Underground Mining Equipment Maintenance

Table 25-3 summarizes the major types of representative underground maintenance errors identified during this project. A number of factors contributing to maintenance-related human error were also identified including:

- Confined work spaces and crowded equipment bays
- Inability to make visual inspections
- Inaccessible components, such as lube points that could not be reached,

adjustment points that are hard to access, major components that could not be reached

- Poor layout of components in a compartment
- Inappropriate placement of components on machines
- Poor or no provision for hose and cable management
- Lack of troubleshooting guides and tools
- Lack of positive component installation guide pins and other installation controls
- Insufficient task inspection and checkout time
- Cumbersome or inadequate manuals
- Excessive weight of components being manually handled

Table 25-3. Typical Mining Equipment Maintenance Errors.[1]

Frequency	Type of Error
I	Install incorrect component
S	Omitting a component
S	Parts installed backwards
S	Failure to properly torque
S	Failure to align, check, calibrate
S	Use of incorrect fluids, lubricants, greases
O	Reassemble error
O	Failure to seal or close
O	Error resulting from failure to complete task due to shift change
O	Failure to detect while inspecting
O	Failure to lubricate
O	Failure to act on indicators of problems due to workload, priorities, or time constraints
O	Failure to follow prescribed instructions

I = Infrequent (less than once per year)
S = Somewhat frequently (2 to 5 times per year)
O = Often (over 5 times per year)

If maintainer-induced errors could be reduced by 50%, overall equipment availability would be increased by more than 10%. These reductions can be achieved through improved design. Listed below are several engineering design improvements that could reduce maintenance errors.

- Improved component-machine interface, including design interface so the component can only be installed correctly, and provide counting pins and other devices to support components while they are being bolted or unbolted
- Improved fault isolation design by designating test points and procedures, providing built-in test capability, and clearly indicating direction of the fault
- Improved indicators, warning devices and readouts to minimize human decision making
- Use of operational interlocks so that subsystems cannot be activated if they are incorrectly assembled or installed.
- Use of positive decision guides to minimize human guess work, for example, arrows to indicate direction of flow, correct type fluids or lubricants, and correct hydraulic pressures
- Design to facilitate detection of errors by locating connections in front of component to facilitate visual inspections, and laying out equipment in a logical flow sequence

STUDY CONCLUSIONS

The following conclusions were derived from the US Bureau of Mines study:

1. There is no evidence of the systematic application of maintainability design principles, concepts, or criteria to the design of operational underground coal mining equipment.
2. Similarly, there is no evidence of systematic application of human factors engineering principles, concepts, or criteria being applied to the design of this equipment with respect to maintenance.
3. Reduced task completion times and fewer maintenance problems were reported for the 10 most frequently performed maintenance tasks on older and smaller machines than for newer more complex equipment. This appears to be the result of simpler design on the older equipment.
4. Increased task complexity and completion times were generally reported for the newer, larger mining machines. This appears to be the result of increased design complexity, larger and heavier components, overlaying of safety and environmental control systems over the basic machine design, and inadequate accessibility to components.
5. For certain machines, heavy maintenance tasks could be performed on the surface or in high roof underground shops equipped with requisite lifting devices. The same maintenance tasks were extremely difficult, time consuming, and risky to perform at the mine face, where they often have to be completed.
6. With the exception of machines produced by one small mining equipment manufacturer, maintenance task completion times for the 10 most

frequently performed maintenance tasks could be reduced from 10% to 30% or more with relatively simple design improvements.

7. Application of accepted human engineering design standards and criteria could substantially reduce maintenance risk. Over one-third of the reviewed maintenance lost-time injuries were traceable to equipment design deficiencies. Estimates of actual maintenance risk reduction resulting from redesign of the equipment could not be derived from the data.

References

1. Conway EJ and Unger R. *Maintainability Design of Underground Mining Equipment, Volume I.* Pittsburgh, PA: US Bureau of Mines; 1985; Final Technical Report. Contract J0145034, VRC Corporation.

2. Conway EJ and Unger R. *Maintainability Design of Underground Mining Equipment, Volume II - Maintainability Design Guidelines.* Pittsburgh, PA: US Bureau of Mines; 1988; Contract J0145034, VRC Corporation, p. 188.

3. Ferguson CA, et. al. *Ergonomics of the Maintenance of Mining Equipment.* Edinburgh: Institute of Occupational Medicine; 1985; 55.

4. Mine Enforcement and Safety Administration. *Analysis of Injuries Associated With Maintenance and Repair in Metal and Non-Metal Mines.* US Department of the Interior; 1985; MESA Information Report No 1058, p. 34.

5. Conway EJ and Sanders M. *Recommendations for Human Factors Research and Development Projects in Surface Mining.* Canyon Research Group, Inc.; 1982; Contract J0395080, BuMines OFR211-83, NTIS PB 84-143650, p 34.

6. Foster J, Phillips D, and Rodgers T. *Reliability, Availability and Maintainability.* M/A Press; 1981; 265.

7. Bird Engineering - Research Associates, Inc. *Maintainability Engineering Handbook.* Naval Ordnance Systems Command Contract N00017-68-4403, 1069, p. 310.

8. Dhillon BS. *Human Reliability.* Pergamon Press; 1986; 239.

9. Dhillon BS and Reiche H. *Reliability and Maintainability Management.* Van Nostrand and Reinhold Publishers; 1985; 240.

10. Conway E and Elliott W. *Mine Maintenance Material Handling: Volume I.* Pittsburgh, PA: US Bureau of Mines; 1988; Final Technical Report. Contract H0113018, Canyon Research Group, Inc.; p. 44.

11. Taylor RJ. *An Introduction to Error Analysis.* Oxford University Press; 1982; p. 270.

12. Christensen JM, Howard JH, and Stevens BS. Field Experience in Maintenance. In Rasmussen J and Rouse WB, eds. *Human Detection and Diagnosis of System Failures.* Plenum Press; 1981; 363.

13. *Effectiveness of US Forces Can Be Increased Through Improved Weapon System Design.* US Government Accounting Office. GAO Report to Congress Number PSAD-81-17. 1981, p 23.

14. Rasmussen J and Rouse W, eds. *Diagnosis of System Failures.* Plenum Press, 1981, 363 pp.

15. Winlund ES and Thomas CS. *Reliability and Maintainability Training Handbook.* Contract NOBs-90331; General Dynamics/Astronautics Division; NAVSHIPS 0900-0023000; 1965; p. 432.

16. Rouse W and Boff K, eds. *Systems Design: Behavioral Perspectives on Designers, Tools, and Organizations.* North Holland Publishers; 1987; 353.

E. J. Kirk Conway is Vice President of VRC Corporation. Mr. Conway has a BS in Psychology. He has seventeen years of experience conducting applied human factors engineering research in military and industrial settings. Mr. Conway has developed equipment and systems design specifications, human engineering specifications and recommended practices, and maintainability design guidelines.

Richard L. Unger is a Project Manager with the US Bureau of Mines specializing in research on the human factors aspects of machine design and materials handling. Some of Mr. Unger's work includes, guidelines for lifting in restricted postures, development of materials handling devices for underground mines, off-road mobile equipment seating design, development of software for the analysis of crewstation designs, and advanced supply handling systems for underground mines. His education includes degrees in the engineering and computer science fields.